W9-DBW-599

MONOGRAPHS OF THE
SOCIETY FOR RESEARCH IN
CHILD DEVELOPMENT

Serial No. 250, Vol. 62, No. 2, 1997

INFORMATION PROCESSING THROUGH THE FIRST YEAR OF LIFE: A LONGITUDINAL STUDY USING THE VISUAL EXPECTATION PARADIGM

Richard L. Canfield
Elliott G. Smith
Michael P. Brezsnyak
Kyle L. Snow

WITH COMMENTARY BY

Richard N. Aslin
Marshall M. Haith
Tara S. Wass
Scott A. Adler

MONOGRAPHS OF THE SOCIETY FOR RESEARCH IN CHILD DEVELOPMENT

Serial No. 250, Vol. 62, No. 2, 1997

CONTENTS

ABSTRACT   v

I. INTRODUCTION   1

II. BACKGROUND AND SIGNIFICANCE   6

III. AGE CHANGES IN ANTICIPATION AND REACTION TIME:
A COMPILATION OF PUBLISHED RESEARCH   18

IV. DESCRIPTION OF THE STUDY AND METHOD   25

V. IDENTIFYING MINIMUM REACTION TIME   36

VI. DEVELOPMENTAL FUNCTIONS AND
INDIVIDUAL DIFFERENCES IN REACTION TIME   53

VII. DEVELOPMENTAL FUNCTIONS AND INDIVIDUAL DIFFERENCES
IN REACTION-TIME VARIABILITY AND ANTICIPATION   93

VIII. CONCLUSIONS AND FUTURE DIRECTIONS   120

REFERENCES   138

ACKNOWLEDGMENTS   145

COMMENTARY

MODELS OF OCULOMOTOR VARIABILITY IN INFANCY
Richard N. Aslin   146

INFANT VISUAL EXPECTATIONS: ADVANCES AND ISSUES
Marshall M. Haith, Tara S. Wass, and Scott A. Adler   150

CONTRIBUTORS   161

STATEMENT OF
EDITORIAL POLICY   163

ABSTRACT

CANFIELD, RICHARD L.; SMITH, ELLIOTT G.; BREZSNYAK, MICHAEL P.; and SNOW, KYLE L. Information Processing through the First Year of Life: A Longitudinal Study Using the Visual Expectation Paradigm. With Commentary by RICHARD N. ASLIN and by MARSHALL M. HAITH, TARA S. WASS, and SCOTT A. ADLER. *Monographs of the Society for Research in Child Development*, 1997, **62**(2, Serial No. 250).

This *Monograph* uses a developmental function approach to describe age-related change and individual differences in infant information processing during the first year of life. The Visual Expectation Paradigm (VExP) is used to measure speed of information processing, response variability, and expectancy formation. Eye-movement reaction times and anticipatory saccades were gathered from 13 infants assessed monthly from 2 to 9 months and then again at 12 months. Analysis of response patterns demonstrated the applicability of the paradigm throughout the age range studied. Converging operations strongly indicate that the traditional estimate of the minimum time required for infants to initiate a saccade to a peripheral stimulus may be as much as 100 milliseconds (ms) too long. Moreover, the newly estimated minimum of 133 ms does not appear to change during the 2–12-month period. Reanalysis of the present data and past research reveals that the new, shorter minimum reaction time is unlikely to affect findings based on mean reaction time. However, using the traditional minimum reaction time will inflate estimates of percentage anticipation, especially in infants older than 5 months.

Group and individual growth curves are described through quantitative models of four variables: reaction time, standard deviation of reaction time, percentage anticipation, and anticipation latency. Developmental change in reaction time was best described by an asymptotic exponential function, and evidence for a local asymptote during infancy is presented. Variability in reaction time was found to decline with age, independent of mean reaction time, and was best described by a polynomial function with linear and quadratic terms. Anticipation showed little lawful change during any portion of the age span, but latency to anticipate declined linearly throughout the first year.

Stability of individual differences was strong between consecutive assessments of mean reaction time. For nonconsecutive assessments, stability was found only for the 6–12-month period. Month-to-month stability was inconsistent for reaction-time variability and weak for both anticipation measures. Analyses of individual differences in growth curves were carried out using random regressions for the polynomial models. The only significant individual difference (in growth curves) was found for reaction-time variability. Parameter estimates from the exponential models for reaction time suggested two or three developmental patterns with different exponential trajectories. This finding indicates that the strong form of the exponential growth hypothesis, which states that processing speed develops at the same rate for all individuals, does not hold for the first year of life.

In the concluding chapter, Grice's Variable Criterion Model (Grice, 1968) is used to integrate three key findings: regular age changes in mean reaction time and variability but no age change in the minimum reaction time. It is argued that the rate of growth of sensory-detection information is developmentally constant during much of the first year but that age changes occur in the level and spread of the distribution of response threshold values. The unique strengths of the paradigm are discussed, and future directions are suggested for further developing the paradigm itself and for using it as a tool to study broad issues in infant cognition.

I. INTRODUCTION

Advances in the scientific study of infant learning and cognitive growth have proceeded primarily through the development of standardized experimental and observational paradigms for assessing fundamental processes. These paradigms typically support two related lines of investigation: investigations into the development of the cognitive process itself and investigations designed to standardize the paradigm. As a result, researchers often find themselves studying their paradigm at the same time they study infant cognition. The research reported in this *Monograph* represents just such a dual focus. Our primary focus is on describing and understanding the development of information processing in the first year of life. However, because we carry out this research using the recently introduced Visual Expectation Paradigm (VExP), we also focus on evaluating the paradigm itself. Our goal is to study age-related change in information processing, but, because age changes in the infant may interact with how the paradigm is implemented, our two foci are far from independent.

Poets and scientists alike have seen the eyes as windows to the mind, and nowhere have the eyes been more important for understanding the mind than in the work of developmental psychologists studying infant perception and cognition. Although human infants lack fine control over their reaching and grasping for many months, their visuomotor system is functional from birth. Developmental psychologists have capitalized on the young infant's visual precocity and devised both global and precise measures of the direction of gaze and the movement of the line of sight as dependent measures in studies of the detection, discrimination, and recognition of stimuli.

Early work by Fantz (1961) introduced the two-choice visual preference method for studying infant perceptual discrimination. Fantz used the simple fact that, when presented with a pair of visual stimuli, even newborns will tend to look at one stimulus more than another. He interpreted an infant's reliable preference for one of a pair of stimuli to indicate a capacity to discriminate between the two. This simple but elegant insight has been the methodological basis for more studies of infant perception and cognition

1

than any other single paradigm and has resulted in carefully standardized habituation-dishabituation of looking paradigms (Bornstein, 1985; Horowitz, Paden, Bhana, & Self, 1972) and response-to-novelty paradigms (Cohen & Gelber, 1975; Fagan, 1970).

The most popular fixation-time paradigms all rely on global assessments of the infant's direction of gaze, usually to a stimulus appearing on either the left or the right of visual center. But, as noted by Aslin (1985), an interpretational difficulty arises when explaining null results. When the infant fails to demonstrate a preference, it may be for any of several reasons. The infant either is unable to discriminate between the stimuli (or detect the presence of the stimulus), is able to make the discrimination but prefers the stimuli equally, or, even though discrimination is possible, does not fixate and encode the critical features of the stimuli that would enable the discrimination to be made.

Partially in response to ambiguities inherent in interpreting relative fixation duration to whole stimuli, in the 1960s researchers began developing methods to measure the precise locations on the stimulus where infants were looking (Kessen, Salapatek, & Haith, 1972; Salapatek & Kessen, 1966). Using infrared corneal reflection photography, these researchers sought to know exactly where on the stimulus infants were looking and for how long they fixated, thereby achieving a more precise and dynamic view of infant visual perception and action. Compared to global methods of relative fixation duration, which sometimes carried the implicit assumption that infant looking was determined by a preexisting preference hierarchy, the scanning methods were able to address the question of *how* infants organize their looking in response to particular stimuli on a moment-by-moment basis (Haith, 1980).

Infant scanning of stationary targets is composed of fixations interrupted by saccades that move the eye to a new fixation point. Saccades are rapid movements of the eyes that serve to place the fovea, the area of the retina most densely packed with photoreceptors (and hence affording the highest visual acuity), in line with the target of interest in visual space. Early studies of infants' scanning of stationary targets addressed several basic issues: (1) Do infants reliably direct their saccades to peripheral targets? (2) Do infants scan entire figures or only a small region of the stimulus? (3) What aspects of infant saccades change with age? Other sources provide excellent reviews of this body of research (Aslin, 1985; Shea, 1992).

Understandably, the focus of early research was on infant saccades to presently visible targets—in the context of either scanning a figure or saccades elicited by the abrupt onset of a peripheral stimulus. It was Haith's fortuitous observation that young infants make saccades to future stimulus locations, even in the absence of current visual information, that led to the development of a paradigm in which both elicited and anticipatory saccades were studied in the same context (Haith, Hazan, & Goodman, 1988; Haith, Wentworth, & Canfield, 1993).

The Visual Expectation Paradigm (VExP) was designed as a dynamic perception-action task in which peripheral stimuli appear on the left and right sides of the infant's visual field in either a predictable or an unpredictable manner. This technique gives infants the opportunity to make elicited saccades in reaction to the onset of a peripheral stimulus, but they are also free to make anticipatory saccades to the location where a stimulus will appear but is not yet visible. Initial research using the paradigm has demonstrated its usefulness for addressing questions about infants' abilities to develop visual expectancies for future events (Haith et al., 1988), simple rule learning (Canfield & Haith, 1991), memory (Wentworth & Haith, 1992), numerical appreciation of sequentially presented items (Canfield & Smith, 1996), stability of individual differences (Canfield, Wilken, Schmerl, & Smith, 1995; Haith & McCarty, 1990), dimensions of infant information processing (Jacobson et al., 1992), and the prediction of childhood intelligence (Benson, Cherny, Haith, & Fulker, 1993; DiLalla et al., 1990). Given its application to such a diverse range of cognitive phenomena, the VExP has the potential to become another standard tool for studying infant cognition and development.

The primary task of developmental psychology is to describe and explain age-related change, but studies of change itself constitute only a small amount of developmental research. Instead, researchers tend to document capacities and stability in those capacities (Appelbaum & McCall, 1983; McCall, Eichorn, & Hogarty, 1977; Wohlwill, 1973). Understanding development requires an understanding of both continuity and change, and to date there have been no published VExP studies describing normative developmental change or documenting individual differences in the nature of change in the basic processes tapped by the VExP, that is, reaction time (RT) and anticipation. Although use of the paradigm is leading to new insights and knowledge about infant information processing and developmental stability, it has so far yielded much less in the way of knowledge about developmental change.

In this *Monograph,* we present our findings from a longitudinal study using the VExP. We adopt a developmental function approach, which leads us to describe, for individual infants, changes in information processing during the 2–12-month age period. This approach gives us the opportunity to study individual patterns of growth and to reveal both continuity and change in the context of a single investigation. Five questions of particular importance are addressed: (1) Are there consistent age changes in RT and the rate of anticipation over the first year of life? If so, what are the shapes of the respective developmental functions? (2) Does a single growth function adequately describe the development of all infants, or are there prototypical growth functions for specific subgroups? (3) Does a single growth function adequately describe development from infancy into adulthood, or does growth appear to take different forms during different periods of development? (4) Is there evidence for short- or long-term stability of individual differences in reaction time and anticipation? (5) In addition to traditional measures of average reaction time (RT) and percentage of stimuli anticipated (%ANT), are there other performance variables from the VExP that reveal other dimensions of infant information processing?

Because the VExP is relatively new, it does not yet possess the degree of standardization enjoyed by other major infant research paradigms. One goal of this *Monograph* is to initiate a dialogue about the paradigm itself and to report research findings that will lead to greater standardization of methods and data handling across ages and laboratories. Thus, the findings reported here have important implications both for future research employing the VExP and for the interpretation of previous research findings.

For example, we propose two new methods whose convergence leads to an empirical determination of the minimum latency at which reactions can occur; these methods have broad consequences for the accurate classification of saccades as reactive or anticipatory. Our application of these methods leads us to propose that RTs can be much faster than the currently accepted value of greater than 200 milliseconds (ms). Moreover, we show that, by defining saccades occurring as late as 200 ms as anticipatory, estimates of %ANT are inflated, especially for older infants.

Finally, we believe that RT is a behavioral index that can be measured throughout the life span. Consequently, many sophisticated models of adult RT may become important resources for theory development and model testing with data from earlier developmental periods. We draw on one of these models, the Variable Criterion Model (Grice, 1968), to address the question, What develops? Taken as a whole, the research reported in this *Monograph* provides a detailed picture of continuity and change in anticipation and RT throughout the first year of life as well as essential information regarding the use of the VExP as a paradigm for studying development.

The plan of the *Monograph* is as follows. In Chapter II, we introduce the conceptual and empirical background by describing the developmental function approach in relation to the VExP. In that context, we define four performance variables and describe the evidence suggesting that they represent significant dimensions of development and individual difference. In Chapter III, we report our findings from an exhaustive compilation of the RT and anticipation data from all published studies that have used an elicited-saccade task with infants. Using these data, we construct population growth curves for average saccade RT and %ANT. An analysis of the limitations of these data leads us to report our longitudinal study of information processing using the VExP. In Chapter IV, we describe the characteristics of the sample and the design, methods, and procedures used. In addition, we evaluate the quality of the data as a function of the age of assessment. In Chapter V, we provide the data analyses needed to determine the minimum possible RT and whether it changes with age, a prerequisite to accurately classifying saccade latencies as either stimulus elicited or anticipatory.

In Chapter VI, we describe age changes and individual differences in RT, including a mathematical description of individual growth curves. In Chapter VII, we present analyses for percentage anticipation and two other performance variables: trial-to-trial variability in RT (SDRT) and latency of anticipations (ANTL). In the first section of Chapter VIII, we assess the effects of using our new, shorter estimate of the minimum RT by comparing age trends for RT and %ANT in the present study using both estimates as well as by reanalyzing key findings from two of our previously reported studies (Canfield et al., 1995; Canfield & Smith, 1996). We conclude that %ANT is highly sensitive to the choice of minimum RT, especially for older infants. In the second section of Chapter VIII, we address the implications of our growth-curve modeling for life-span theories of the nature of growth in processing speed (Hale, 1990; Kail, 1988, 1993b). In the third section, we use the Variable Criterion Model (Grice, 1968) to address the question of which aspects of processing speed develop during the first year of life. Finally, we discuss what we see as the most important challenges facing the VExP as well as promising directions for future research.

II. BACKGROUND AND SIGNIFICANCE

The VExP arose from Haith's serendipitous observations—made over a period of years studying infant visual perception, cognition, and oculomotor control—that young babies sometimes appear to make anticipatory visual fixation shifts (e.g., Haith, 1980; for a historical account of the paradigm's origins, see Haith et al., 1993). The Visual Expectation Paradigm (VExP), which Haith developed to reveal these anticipations, was introduced into the research literature in 1988 (Haith et al., 1988). In the VExP, an infant typically views a sequence of visual stimuli that appear for a brief time to the right or left of visual center. As the baby views the sequence of pictures in the dark, infrared photography is used to record eye movements onto videotape. Tapes are then coded off-line to identify saccade latencies for each picture presentation.

The VExP is founded on a distinction between two types of saccades. A *reactive* saccade occurs when the baby detects a peripheral stimulus and makes an abrupt shift in visual fixation to align the fovea with the visual target. A reactive saccade is therefore elicited by parafoveal stimulus information. In contrast, *anticipatory* saccades are guided by information in the baby's memory about the likely time of occurrence and spatial location of a future stimulus (Smith, 1995). Anticipatory saccades to a predicted location are initiated before a target is displayed or so quickly after stimulus presentation that the internal command to move the eye must have been generated before the time of picture onset. Functionally, reactive and anticipatory saccades are distinguished by a latency criterion. The fastest possible reaction latency represents an irreducible minimum RT, and a fixation shift to a stimulus location prior to this minimum latency is classified as anticipatory.

Although the empirical research using the VExP has addressed various questions about infant cognition, there has been little focus on the nature of age differences in RT and %ANT. For example, although the range of infant ages in published studies is quite large (2–8 months), individual investigations have rarely included infants from more than a single age group. Only one published study (Canfield et al., 1995) has employed a longitudinal methodology. Thus, although the paradigm has shown considerable promise as a tool for understanding infant cognition and information processing, there does not yet exist a substantial research base for evaluating its use as a tool for investigating infant *development.*

A DEVELOPMENTAL FUNCTION APPROACH

Research on infant information processing using the VExP can become more developmental by using a developmental function approach to focus on changes and continuities in behavior (Appelbaum & McCall, 1983; Burchinal & Appelbaum, 1991; Wohlwill, 1970, 1973). A developmental function describes the relation between age and an individual's response on a specified dimension of behavior over some significant portion of the life span. Describing the form of this relation for individuals, its commonality across individuals, and its sensitivity to environmental variations is often considered to be the primary task of the developmental psychologist (Wohlwill, 1970, 1973).

The first step in a developmental function analysis is to specify the dimension along which behavioral change is thought to occur. Once the dimension is specified and appropriate measures identified, the form of the developmental function can be described in several ways. For variables measured on quantitative scales, one can describe the function in terms of monotonicity or nonmonotonicity, note the presence and general direction of change, note the minimum and maximum values and any inflection points, and even specify the form of the function in terms of either the general family of mathematical functions (e.g., polynomial, exponential) or a specific function having particular parameter values (Burchinal & Appelbaum, 1991; Kail, 1991c). The commonality across individuals concerns the degree to which particular groups of individuals having the same or similar functions can be identified.

The most basic goal of the developmental function method is to produce a detailed description of the pattern of age-related change. Ultimately, this methodology can be used to explain development in terms of how particular environmental and genetic variations affect the nature of growth along specific dimensions—a goal often achieved in studies of physical growth (e.g., Bogin, 1988) but infrequently achieved in psychological studies (Appelbaum & McCall, 1983). In the case of cognitive development, one may seek to describe the effects of an enriched or impoverished caregiving environment on the developmental function relating age to vocabulary size or the relation between mother's education and the pattern of age changes in infant information processing (e.g., Colombo & Mitchell, 1990).

The descriptive focus of the developmental function approach is sometimes avoided by developmental psychologists who have worked hard to help their profession shed the stigma, earned early in this century, of being a purely descriptive enterprise (Wohlwill, 1973). Nevertheless, a solid base of descriptive information remains the appropriate foundation for explanation and theory development. As Wohlwill (1970) observed, "In any area concerned with the study of change, the place of descriptive study looms particularly large, since a complete picture of the course of change will frequently convey some direct information concerning the processes which govern that change" (p. 176). Three fundamental features distinguish mere atheoretical description from description that leads to an explanation of development: (1) incorporation of age into the definition of the dependent variable; (2) a focus on describing age changes for specific individuals; and (3) the selection of an appropriate behavioral dimension along which change is measured.

Use of the age variable has frequently been criticized because of the lack of prediction it affords regarding an individual's behavior (Kessen, 1960). At any given age, behavioral tendencies and capacities vary widely from one individual to the next. But, when explaining the pattern of change *itself* becomes the focus of the investigation, the age variable is no longer an independent variable. The question changes from what age predicts to what factors influence the way behavior changes as a function of age; that is, the question becomes, What influences the shape of the developmental function?[1]

If descriptions of age changes are to result in explanations of development, then it is essential that growth be described at the level of individuals. Descriptions of age changes based on cross-sectional data are inadequate for addressing questions about individual differences in the pattern of change. Without longitudinal data, one embraces the usually questionable assumption that the nature of growth is the same for all individuals. Comparing developmental functions for different individuals will yield information about universal as well as individual change. In the end, explanation is facilitated by longitudinal research because patterns of individual differences in growth provide an additional source of information for identifying the causes and correlates of particular growth functions.

The most important factor determining the explanatory value of a description is the choice of the developmental dimension along which change is described. A developmental dimension is a special case of a scientific construct, that is, an abstract quality of behavior that cannot be directly observed or measured (Nunnally, 1978). When psychologists measure behavior, they are rarely interested in the concrete observable properties of the response itself. Rather, they usually care about some more abstract quality that the response is assumed to reflect and that has meaningful relations to other

---

[1] For an extensive discussion of the status of the age variable in developmental research, see Wohlwill (1973).

constructs. For example, although reaction time can be very accurately measured, one is not normally interested in merely how fast an individual responds. Instead, one assumes that response speed reflects some broader, more abstract quality of behavior (e.g., processing speed, encoding speed, or decision speed). Furthermore, these constructs may be thought to relate in a meaningful way to an even broader attribute, such as information-processing ability. Thus, "a construct represents a hypothesis (usually only half-formed) that a variety of behaviors will correlate with one another in studies of individual differences and/or will be similarly affected by experimental treatments" (Nunnally, 1978, p. 96).

A developmental dimension is a special case of a construct that refers only to those abstract qualities of behavior that show lawful changes with age. To be considered as a potential measure of a developmental dimension, not only must a variable be lawfully related to other supposed measures of the same construct, but it must also demonstrate lawful age changes during a significant portion of the life span. Therefore, construct validation in the developmental sciences involves both horizontal (across measures) and vertical (across age levels) components. The identification and validation of theoretically meaningful or practically useful constructs is a primary task for all the sciences, and developmental science is no exception. However, the nature of developmental constructs demands additional constraints. Unlike a generic psychological construct, whose label must be meaningful in relation to a theory about the nature of behavior, a developmental construct must also be meaningful in relation to a theory about the nature of the growth process.

In this *Monograph,* our exploration of four variables measured in the VExP represents a very early attempt at construct identification and validation. We propose that these variables may correspond to three or more dimensions of infant information processing that have developmental significance. We label these dimensions *speed of information processing,* which is measured by reaction time (RT) and anticipatory saccade latency (ANTL); *variability in processing speed,* which is measured by trial-to-trial variation in RT (SDRT); and *expectancy formation,* which is measured by the percentage of pictures anticipated (%ANT). The amount and strength of evidence supporting the appropriateness of these labels vary considerably from one variable to the next. Although labeling the dimensions may be considered premature in light of the paradigm's relative youth, the identification of variables and their relations to broader dimensions of significant developmental change must be pursued in a parallel manner with the description of age changes in those variables. The validation of a putative developmental dimension depends not only on identifying multiple measures of a given construct but also on the degree to which age-related change proves lawful.

Lawfulness may mean simply that regular age changes are found to occur

9

in the variable, but it also often implies that some continuity is found in individual performance over time. The concept of continuity in development is often identified only with findings of stability of individual differences across age. However, as Emmerich (1964) and others have emphasized, continuity in development has two largely independent referents. One notion of continuity concerns the form of the developmental function that relates changes in behavior to changes in age, typically, whether it is continuous or discontinuous with respect to the underlying process being measured. (Continuous functions are used to represent development that is primarily quantitative and cumulative in nature; but, if the attribute being studied undergoes qualitative shifts, the underlying function is typically conceptualized as discontinuous.)

Independent of the nature of the developmental function, continuities can also be found in the realm of individual differences. This form of continuity, which Emmerich termed *stability,* concerns the degree to which individuals maintain their rank order on the same measure over time. Development with respect to a particular dimension is often considered lawful when stability of individual differences is found, at least for portions of the growth curve. Without at least modest stability, it is difficult to know if one is capturing anything systematic about development. However, as Appelbaum and McCall (1983) have cautioned, a lack of longitudinal stability may simply reflect the presence of developmental change, and the utility and validity of a measure may still be revealed by its correlates at a specific age.

Although, at the time of this writing, fewer than ten studies using the VExP have appeared in the literature, those studies indicate that average RT and %ANT (*a*) have lawful relations with broad constructs of information processing, (*b*) show lawful relations to variables from other infant paradigms that are also thought to measure these dimensions, and (*c*) have both short- and long-term developmental stability. In the next section, we assess in more detail the accrued evidence from these studies supporting the validity of the constructs we postulate. Taking each measure in turn, we present the evidence and rationale for evaluating the four measures from the VExP as candidates for charting the course of cognitive growth.

Saccade Reaction Time

In the VExP, infants are exposed to colorful moving images. If the infants are quick enough, they can foveate the images for several hundred milliseconds by responding with a reactive saccade. For a given stimulus presentation, the time from the onset of the stimulus to the beginning of eye rotation (initiated after the minimum RT) provides a single RT measure for that trial. The mean or median of a set of RT values over a specific type and number of trials is typically used to represent the average RT for each infant. The number of values entering into an average RT ranges from about five for *baseline* RT to 50 or more for *postbaseline* RT. Because the VExP has been implemented as a free-looking procedure,[2] the absolute number of RTs entering into an average RT depends on two factors: the frequency of off-task behaviors and the number of anticipatory looks. Infants who engage in more off-task behavior (e.g., by closing their eyes or otherwise looking away from the stimulus screen) will typically have fewer RT values. Similarly, infants who anticipate a greater proportion of pictures will necessarily react to fewer.

Early studies sought evidence that infant RT measured by saccade latency was related to cognitive processing in some lawful manner. Haith et al. (1988), for example, reported that RTs in 3.5-month-old infants were shorter when stimuli appeared in a predictable as opposed to an unpredictable sequence. Following this, Canfield (1988; Canfield & Haith, 1991) reported that RTs were faster at 3 than at 2 months of age and that the amount of decline in RT from baseline to postbaseline increased linearly as a function of the complexity of the stimulus sequence for 3-month-olds.

More recent work has addressed the question of whether RT is a reliable measure of individual differences. One study demonstrated that RT shows significant test-retest reliability over a span of several days for 3-month-olds (Haith & McCarty, 1990), and another reported that average RT is very stable from 4 to 6 months (Canfield et al., 1995). Two additional investigations examined the predictive validity of the VExP. DiLalla et al. (1990) reported a significant correlation of $-.47$ between infants' baseline RT measured at 8 months and their 3-year Stanford-Binet IQ. Benson et al. (1993) reported a significant correlation of $-.28$ between *mid-twin* baseline saccade RT and *mid-parent* WAIS scores.

Finally, a study employing multiple measures of infant information processing in babies 6.5–12 months old suggests that average RT represents one measure from a larger set of measures all of which reflect a coherent dimension of behavior, a dimension labeled *processing speed* (Jacobson et al., 1992). Jacobson and her colleagues tested 6.5-month-old infants using the VExP and a visual recognition memory (VRM) task originated by Fagan and Singer (1983). Then at 12 months the infants completed a second VRM task and a cross-modal transfer task (Rose & Feldman, 1987; Rose, Feldman, & Wallace, 1988). The VRM and cross-modal tasks each provided estimates of fixation duration, an accepted measure of infant processing speed (Colombo, 1993; Colombo, Mitchell, O'Brien, & Horowitz, 1987).

---

[2] A free-looking procedure has also been used in habituation studies where the stimulus is continuously available and babies accumulate looking time from their spontaneous (not elicited) looks at the stimulus (see Bornstein, 1985). Similarly, in the VExP, babies are presented with a sequence of stimuli that they view as often or as little as they wish.

When VRM fixation duration was averaged for the 6.5- and 12-month testing, positive and significant correlations were obtained between fixation duration and both baseline and postbaseline RT (r's = .24 and .28, respectively). In addition, 12-month fixation duration in the cross-modal task was positively and significantly correlated with both RT measures (r = .33 for both). Most important, when the entire set of information-processing measures was submitted to factor analysis, the RT and fixation duration measures all loaded on the same factor. Moreover, the processing-speed factor was distinct from two other factors, whose loadings indicated that they measured somewhat different dimensions.

Although they loaded on the same factor in the Jacobson et al. (1992) study, bivariate correlations indicate that RT and fixation duration share only about 10% variance. Furthermore, a rational task analysis suggests substantial independence between the two measures. Encoding speed has been proposed as a candidate for the process underlying fixation duration (Colombo, 1993; Fagan, 1984), but encoding the details of a visual stimulus is not typically called for in the VExP (but see Wentworth & Haith, 1992). In fact, because the visual images are typically randomized and appear for less than a second in peripheral vision, the infant has little opportunity and presumed motivation to encode information about form or other details of the stimulus. Instead, RT in the VExP would seem to reflect more the processes of detecting peripheral stimuli, deciding what action to take, and programming the appropriate response. Recently, Bertenthal (1996) has proposed the existence of two functionally dissociable perceptual systems in early development. One system is concerned with perception for object recognition and the other with perception for the guidance of actions. The Jacobson et al. (1992) research suggests that some of the factors influencing speed of processing affect both systems.

It is sometimes thought that stimulus-elicited saccades are reflexive and therefore devoid of cognitive content. However, research indicates that stimulus-elicited saccades are influenced by spatial and temporal predictability, both in adults (Abrams & Jonides, 1988; Findlay, 1981; Michard, Tetard, & Levy-Schoen, 1974) and in infants (Canfield & Haith, 1991; Haith et al., 1988). In addition, the term *reflex* is problematic in that it does not appear to refer to a distinct class of behaviors (Findlay, 1992; He & Kowler, 1989; Zingale & Kowler, 1987).

In summary, saccade RT in infants seems to be a very simple response that relates in lawful ways to the age variable, concurrently to other measures of infant cognition, and predictively to later IQ. Furthermore, its test-retest reliability and cross-age stability suggest that it has the potential to be psychometrically sound. For these reasons, we believe that saccade RT is a promising measure for developmental analysis.

Trial-to-Trial Variability in Saccade Reaction Time

Variability in performance has been explored as a general dimension of individual difference for adults in the realms of cognition, personality, and aging (Eysenck, 1986; Feingold, 1995; Fiske & Rice, 1955; Hale, Myerson, Smith, & Poon, 1988). Therefore, we considered the merits of performance variability as a source of information about the nature of individual differences and developmental change during infancy.

Research on elementary cognitive processes in adult populations suggests that the VExP may offer an additional useful performance measure, namely, trial-to-trial variability in RT. Several studies have found that children's latencies on speed tasks are more variable than those of young adults and that older adults' latencies are more variable than those of younger adults (Hale et al., 1988; Jensen, 1992b; Morse, 1993; Poon, Myerson, Hale, & Smith, 1992; Verhaeghen & Marcoen, 1990). However, these studies also find that average RT (mean or median) and trial-to-trial variability (SDRT or interquartile range of RT) are highly correlated.

Some researchers view SDRT as representing error associated with the measurement of average RT or as being wholly redundant with the mean (Hale et al., 1988; Poon et al., 1992). Others argue that RT variability is an important measure in its own right. For example, Carlson, Jensen, and Widaman (1983) reason that individuals who have lapses of attentiveness when performing a speed task will be more variable, and Eysenck (1986) suggests that RT variability is a result of neurobiological factors. According to Eysenck's theory, SDRT reflects underlying variability in the transmission of nerve impulses. Using this theory, Jensen (1992a, 1992b, 1993) argues that SDRT is a direct contributor to associations between performance on simple speed tasks and individual differences in psychometric g. In support of this argument, Jensen (1992a) reports that associations between SDRT and g are consistently higher than associations between mean RT and g in the same sample. Finally, a study by Larson and Alderton (1990) indicates that the correlation between RT variability and IQ is highest for the slowest segment of the RT distribution. Looked at in this way, SDRT reflects not measurement error but rather a distinct dimension of stimulus processing that is related to the broader construct of intelligence and on which individuals can be found to differ.

No published studies using the VExP have reported trial-to-trial variability, and, although no general agreement exists about its proper interpretation, we believe that research on adults and children suggests its potential validity as an index of a personal attribute relevant to cognitive functioning. If we find changes in SDRT to be lawfully related to age, independent of the mean RT, or if we find stable individual differences in SDRT, it would indi-

13

cate that further study of the dimension of trial-to-trial variability during infancy is warranted.

Expectancy Formation

Most of the early research on anticipation and expectation formation was undertaken within the framework of Piagetian theory. At times, Piaget suggested that anticipation was a defining feature of cognition and that lawful qualitative changes occurred on that dimension. For example, in *Origins of Intelligence* (1952), Piaget describes how the newborn sucks reflexively when stroked on the mouth or cheek but notes that sucking can also be observed in the absence of a stimulus. This led him to see sucking behavior as internally generated—what he labeled a *functional assimilation* of the primitive sucking scheme—rather than as simply evoked by a specific stimulus.

Crucial for Piaget's theory was the separation between behaviors and an eliciting stimulus. He described how early functional assimilations gradually develop into the primary circular reactions of Sensorimotor Stage 2, and he identified his son Laurent's anticipatory sucking when held in the nursing position as the first instance of anticipatory behavior, appearing when he was about 2 months old. By the age of 3 months, Laurent was able to discriminate between being held in a nursing position by his father and being held by his mother—engaging in anticipatory sucking only when being held by his mother.[3]

Piaget interpreted these anticipations as indicative of an active mental life, one in which the infant is not a slave to stimuli but rather is self-directing and uses memory as a guide for anticipating the future. These early anticipations were seen to constitute a foundation for the later development of more complex expectancies, such as means-ends relations and object constancy in the latter half of the first year of life (Harris, 1983). Finally, the development of representational thought allows the child to anticipate events and plan actions that are mentally but not perceptually given and, therefore, to engage in true thought (Piaget, 1952).

Piaget's analysis suggested that a situationally appropriate action preceding a target event is evidence for an expectancy. These actions prepare the infant for a specific upcoming event, as when Laurent began sucking before being put to the breast. But Piaget's terminology did not make explicit the

[3] Note that there are some similarities to Piaget's description of age changes in what he called "anticipation" and the findings of Canfield and Haith (1991). Simple anticipation appeared available to the 2-month-old infants, whereas clear evidence of discriminative anticipation was present only for the 3-month-olds. In the left-left-left-right condition, only the older infants demonstrated that they had discriminated the first and second left pictures from the third left picture—showing a higher rate of anticipation after the latter.

distinction between the cognitive construct and the behavioral act. Thus, when Haith et al. (1988) introduced the VExP, they proposed a distinction between the behavioral actions that precede an event, that is, *anticipation,* and the cognitive construct that motivates it, that is, *expectation.* We adopt this terminology here.

An anticipatory eye saccade is one of a possibly larger set of observable behaviors that may indicate a cognitive expectancy. The term *expectation* has also been used to explain surprise reactions and dishabituation to novelty, but in these cases no distinction is made between the violation of an expectancy, indicating that the infant had actually forecasted the appearance of the familiar stimulus, and the initiation of an orienting reaction elicited by the novel stimulus (Canfield & Haith, 1991; Haith et al., 1993). In relation to dishabituation, the anticipatory saccade constitutes less ambiguous behavioral evidence of an expectancy, both because the saccade commences before the stimulus and because it is directed to a specific location.

Under the assumption that %ANT in the VExP reflects the broader construct of expectancy formation, one could reasonably expect that infants will anticipate more as they get older. However, findings of higher %ANT in older babies are not always reported. For example, Canfield and Haith (1991) reported greater anticipation by 3-month-old than 2-month-old infants when viewing asymmetrical sequences (left-left-right and left-left-left-right) but no significant age differences when viewing a simple left-right (L-R) alternating sequence. Similarly, Wentworth and Haith (1992) found no significant age differences for simple alternating sequences. Because these studies were cross-sectional in design, the null results could be the result of high individual variability, and the between-subjects comparisons across age may therefore have lacked the necessary power to reveal age differences. However, in their longitudinal study of infants at 4 and 6 months, Canfield et al. (1995) did not find an increase in anticipation. Instead, they found significantly *lower* rates of anticipation at the 6-month follow-up visit.

In addition to uncertainty about the existence and direction of age changes, whether there is stability of individual differences in %ANT is also uncertain. Haith and McCarty (1990) reported that %ANT showed modest test-retest reliability for 3-month-olds over the span of 3–7 days, but Canfield et al. (1995) found no longer-term stability from 4 to 6 months.

Regardless of inconsistent findings of age changes and stability, predictive validity has been indicated in both the DiLalla et al. (1990) and the Benson et al. (1993) studies. These studies report that %ANT at 8 months (in an unpredictable stimulus condition) correlates significantly with 3-year Stanford-Binet IQ (DiLalla et al., 1990), and mid-twin percentage anticipation was found to correlate significantly with mid-parent WAIS scores in both studies of 8-month-old twins (Benson et al., 1993; DiLalla et al., 1990). Furthermore, some evidence suggests substantial overlap between the dimension

15

of expectancy formation and the broader dimensions of memory and attention. Jacobson et al.'s (1992) study of multiple information-processing measures revealed, for 6.5-month-olds, a significant correlation ($r = .25$) between %ANT and preference for novelty measured by a visual recognition memory task (Fagan & Singer, 1983).

In summary, although a rational task analysis would suggest that a coherent dimension of expectancy formation exists, the empirical evidence is mixed. Regular age changes and stable individual differences in %ANT over a span of months has not yet been demonstrated, but prediction of later ability and coherence with other infant measures suggest that some meaningful dimension of development is being measured. Research using consistent stimuli and procedures across a broad age range is needed to address these questions.

Anticipation Latency

Expectancies are important for facilitating the smooth integration of precise motor actions with a dynamic flow of changing sensory information over which we have no control. Roberts (Roberts, Hayer, & Heron, 1994; Roberts & Ondrejko, 1994) has suggested that the problems associated with articulating perception and action present a flexibility/precision dilemma that is solved by expectancies that enable us to constrain the number of potentially relevant responses to a forecasted event. As a result, we are able to choose an action from a small set of precise responses, each of which is appropriate for a slightly different set of real circumstances. For example, a batter in baseball must be exceptionally precise to hit the ball. At the same time, the batter must maintain the flexibility necessary to hit the fast ball, the curve ball, or the change-up that is thrown on any given pitch.

Expectancies are a necessary part of successful hitting. On the one hand, the batter studies the individual pitcher's repertoire to develop general expectancies regarding what types of pitches are most and least probable. But the batter must also detect arm speed and how the ball leaves the pitcher's hand for a given pitch. The two types of expectancies, those related to a pitcher's repertoire and those related to arm speed and delivery, exist on different time scales. In the first case, the expectancy can be formulated without any relevant time constraints, but, in the second case, the batter must generate the appropriate expectancy about what pitch is coming very quickly, before it is necessary to initiate the anticipatory act of swinging the bat.

The situation facing babies in the VExP is similar in some respects to that of the batter at the plate. In the L-R sequence, for example, when a picture on one side disappears, the baby must activate an expectancy for the picture to shift sides and send the appropriate command to the eye muscles so that the eye begins to rotate. These steps (or their equivalent) must be carried out quickly for the baby to have his or her eyes in the correct position early enough to get a good look at the attractive picture, which would lead to anticipations. Consequently, a limiting factor in making anticipations is not only memory for what is likely to happen but also the ability to activate the appropriate expectancy very quickly. Regardless of whether the baby knows what will happen, if he or she is too slow to activate the appropriate expectancy, then anticipations will be rare. If processing speed increases with age, then we might expect older infants to be capable of activating an expectation earlier and thereby making their anticipations earlier in the interstimulus interval than younger infants. This would be a much less taxing viewing strategy because it affords the baby more control over his or her own visual experience. Thus, even if older babies do not generally anticipate more frequently, we may find evidence for development in anticipation by examining their anticipation latencies.

This task analysis suggests to us that average anticipation latency may represent a second measure of processing speed in the young infant. The process that we believe it reflects, anticipatory saccade programming, is neurally somewhat distinct from reactive saccade programming because it requires a command originating in the frontal eye fields (Hallett & Lightstone, 1976; Smith, 1995; Tusa, 1990).

CONCLUSIONS

The published research from the VExP reveals three possible dimensions of infant information processing: processing speed, expectancy formation, and trial-to-trial variability. Only the first two have been examined in previous studies. Our review of the VExP research reveals that most studies have included infants of a single age or, at most, two ages no more than 2 months apart. The reported within-study age differences provide us with an inconsistent picture of the nature of development. Although it has been found consistently that older infants react faster than younger infants, older infants are rarely found to anticipate more. Nothing has yet been reported about SDRT or ANTL.

Although age differences may not be consistently revealed in one or another particular study, by combining data across many studies the general shapes of the developmental functions for these processes may be revealed. Therefore, as a first attempt to describe age changes in RT and %ANT, it will prove instructive to bring together the findings from studies that have used an elicited-saccade task with infants. We turn to this task in Chapter III.

III. AGE CHANGES IN ANTICIPATION AND REACTION TIME: A COMPILATION OF PUBLISHED RESEARCH

Normative age-related change in average saccade latency is considered briefly in Shea's (1992) overview of research on infant visual development. Shea noted that, between the ages of 1 and 2 months, average saccade latencies declined from about 1,000 ms to about 700 ms (Aslin & Salapatek, 1975). The only other developmental data discussed by Shea are two studies of children's saccades. In one investigation, Cohen and Ross (1977) studied 7–9-year-old children's saccades to horizontally displaced targets. Average latencies at this age were about 270 ms, and adults averaged about 210 ms.[4] In a subsequent study, Groll and Ross (1982) reported slightly longer latencies for 5–6-year-olds than for 8–10-year-olds. These data indicate that, under comparable stimulus conditions, saccade latency declines by about 730 ms between the ages of 1 month and 10 years and then by an additional 60 ms by adulthood. Until recently, little was known about the course of age changes in saccade latency between 2 months and 5 years, but data are now available to describe in some detail age trends in saccade RT. We are aware of no report, however, that has brought the data from various studies together to describe and summarize the nature of these age changes.

ANALYSIS OF THE INFANT SACCADE LATENCY DATA

Selection of the Data Set

The data that we consider derive primarily from research using the free-looking VExP, in which there is typically a 500–1,500-ms no-stimulus interval between stimuli. Data are also available from investigators using experimenter-controlled trials procedures (Aslin & Salapatek, 1975; Hood & Atkinson,

---

[4] Exact mean latencies are not reported because the original publication reports the data only in a figure.

1993; Johnson, Posner, & Rothbart, 1991, 1994). For studies using trials, the infant's point of regard is brought to the center of a display screen by an attractor stimulus, and then a peripheral stimulus is displayed. The attractor may continue to be displayed after target onset, it may be extinguished coincident with target onset, or it may be extinguished prior to target onset. Many of these studies provide measures of %ANT in addition to RT data, although the conditions under which these responses are elicited vary from one study to the next. By collating the data from all these studies, we can begin to describe age changes in RT and %ANT. (None of these studies reports anticipation latencies or variability in RT.)

The data set for the present analysis consists of all data on the latencies of infants' saccades to briefly presented visual stimuli published in major North American journals. Our literature search revealed 12 studies and a total of 44 experimental conditions, with infants ranging in age from 1 to 8 months. The sources and characteristics of these data are shown in Table 1.

Table 1 includes average RTs and %ANTs from all the studies included. Since elicited-saccade tasks were developed for studies of early infant development, the majority of the data points entered for the present analysis come from infants under the age of 5 months, which is the mean age for the entire set of studies. In studies from both the VExP and the trials procedures, viewing conditions vary widely. For some experimental conditions, infants viewed a simple alternating (L-R) sequence, while, in others, they viewed more complex predictable (L-L-R or L-L-L-R) and unpredictable (irregular [IR]) sequences. Some conditions involved prediction rules based on stimulus identity or cardinality/ordinality. In addition to variability associated with sample selection, the differences in sequence complexity and in specific laboratory protocols for collecting and coding data should lead to a range of estimates for mean RT and %ANT for each age group. But, correspondingly, there is no particular experimental condition that represents a generally agreed-on standard for measuring these attributes. We chose to include a single data point for each different age group and condition in a given study (see Table 1). As a result, our conclusions will tend to represent an average level of performance at each age across a number of different stimulus conditions that all produce data on elicited RT or %ANT, or both.

Age Changes in Saccade Reaction Time

Average (mean or median) RT for each condition for all studies in Table 1 is plotted as a function of age in Figure 1. As shown in the figure, there appears to be substantial age-related change. Mean RT declines very rapidly until about 5 or 6 months, followed by little subsequent change through 8 months. In general, latencies in each age group are faster for infants tested

TABLE 1

SUMMARY OF PUBLISHED RESEARCH FINDINGS THAT MEASURE AVERAGE RT AND
PERCENTAGE ANTICIPATION DURING THE FIRST YEAR OF LIFE

Age, Study, and Condition	Average RT	% ANT
1 month		
Aslin & Salapatek (1975):		
10° & 20° combined	1,060[a]	...
1.5 months		
Hood & Atkinson (1993):		
720-ms gap	824	...
2 months		
Aslin & Salapatek (1975):		
10° & 20° combined	700[a]	...
Canfield & Haith (1991, experiment 2):		
IR	647	19.0
L-R	563	27.0
L-L-R	680	20.5
L-L-L-R	552	21.0
Johnson et al. (1991):		
Test trials	...	13.9[b]
Wentworth & Haith (1992):		
Study 1	547	15.1
Study 2	567	16.2
Study 3	526	27.6
3 months		
Canfield & Haith (1991, experiment 2):		
IR	553	24.6
L-R	506	34.0
L-L-R	511	38.0
L-L-L-R	549	36.0
Haith & McCarty (1990):		
Session 1	480	17.4
Session 2	445	19.4
Hood & Atkinson (1993):		
720-ms gap	459	...
Johnson et al. (1991):		
Test trials	...	14.5[b]
Wentworth & Haith (1992):		
Study 1	537	17.2
Study 2	463	20.3
Study 3	503	22.8
3.5 months		
Haith et al. (1988):		
L-R first	409	10.4
L-R second	373	33.7
IR first	478	14.8
IR second	445	7.4
4 months		
Canfield et al. (1995):		
L-R	428	26.7
Johnson et al. (1991):		
Test trials	...	29.2[b]

TABLE 1 (*Continued*)

Age, Study, and Condition	Average RT	% ANT
Johnson et al. (1994):		
Cued training	497[c]	...
Uncued control	474[c]	...
5 months		
Canfield & Smith (1996):		
Experiment 1, IR	378	...
Experiment 1, no. 2	383	...
Experiment 2, IR	375	...
Experiment 2, no. 3	403	...
6 months		
Canfield et al. (1995):		
L-R	347	18.3
Hood & Atkinson (1993):		
720-ms gap	449	...
6.5 months		
Jacobson et al. (1992):		
IR/L-R	323	28.1
8 months		
DiLalla et al. (1990):		
IR	...	17.0[d]
L-R	420[e]	17.0[d]
L-L-R	...	12.0[d]
Benson et al. (1993):		
IR	...	11.9[d]
L-R	360[f]	19.3[d]
L-L-R	...	15.3[d]

NOTE.—Except where indicated, reaction time and percentage anticipation measures reported are based on the total data available (global) or the postbaseline segment of the data.

[a] Includes horizontal and diagonal saccades in the replacement trials.

[b] Infants had already had a training phase when the reported percentage anticipation measure was observed. Only "training trials" in the test sequence are included in the percentage anticipation score.

[c] Based on first 16 trials of session.

[d] Infants viewed a frequently changing sequence within each subject's session, resulting in multiple, nonconsecutive segments of each pattern type within a session. The percentage anticipation measures reported are collapsed across these segments.

[e] Report includes only baseline RT for the first five pictures in the first 1/1 segment.

[f] Report includes only baseline RT for the first eight pictures in the first 1/1 segment.

in the (free-looking) VExP than in the (experimenter-controlled) trials procedure. This is probably due in part to the fact that, in the typical VExP study, stimuli are spatially or temporally predictable. By contrast, in the experimenter-controlled procedure, it is more rare to have temporal predictability, and it is common for the attractor stimulus to overlap with the target, which lengthens saccade latencies (Hood & Atkinson, 1993). Linear regression for the values in Figure 1 indicates a decline in RT of approximately 60 ms/month ($F[1, 34] = 33.51$, $p < .0001$), but the shape of the relation suggests that the true function is nonlinear.

From these data we learn that RT declines systematically as a function

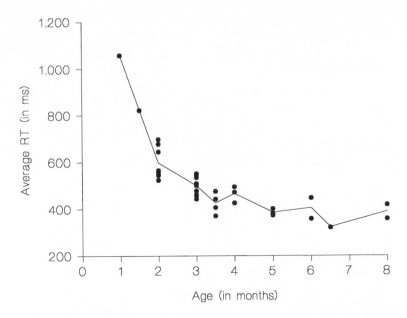

Figure 1.—Plot of average saccade reaction time (RT) as a function of age. Each point represents the average saccade latency for an experimental condition from the studies reviewed. A list of studies included can be found in Table 1.

of age, and we get an estimate of the rate (linear slope) of development in this domain. We also note that the rate of change is not constant across age: development proceeds more rapidly for younger infants.

What we cannot determine is whether the developmental function derived from such a diverse set of studies is representative of the growth in speed for an individual child. We cannot determine whether, given identical protocols, individual infants will show regular age changes that can be similarly described in either a linear or a nonlinear manner. Finally, we cannot determine whether or at what rate RT continues to decline after 8 months.

Age Changes in Anticipation

All but four of the studies cited in Table 1 also report mean percentage of trials anticipated. These data are plotted as a function of age in Figure 2. Averaging across studies within each age level to construct a function line reveals no consistency in the relation of %ANT to age in the 2–8-month period. Not only is there no apparent increase in anticipation rate with age, but it also appears that the oldest age group anticipates the least (see Figure 2). Although there are very few studies of visual anticipation that include two age groups in the same experimental conditions, recall that a finding of less

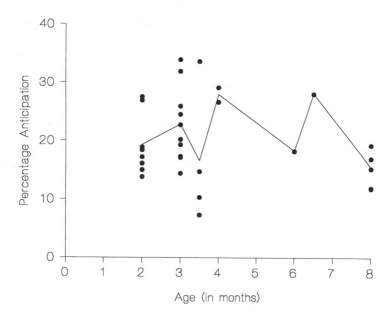

FIGURE 2.—Plot of average percentage visual anticipation as a function of age. Each point represents the average percentage anticipation for an experimental condition from the studies reviewed. A list of studies included can be found in Table 1.

anticipation in older babies is not altogether inconsistent with existing longitudinal research. As previously mentioned, the only longitudinal study that measured %ANT reported a significantly *lower* level of anticipation when infants were 6 months old than when they were 4 months old—18% and 27%, respectively (Canfield et al., 1995).

CONCLUSIONS

By bringing together all the published elicited-saccade latency and anticipation data for infants, we learn that age changes in RT appear to be lawful but that no such regularity is evident for %ANT. This leads to the tentative conclusion that RT but not %ANT represents a potentially useful measure of a coherent dimension of development. However, the growth curve derived from participants from different laboratories, all experiencing a somewhat different implementation of an elicited-saccade task, is an inadequate basis from which to understand development for the individual. In addition, differing implementations of the paradigm across laboratories may be confounded with age. That is, for studies using the VExP, younger infants (2–3.5 months old) have tended to be studied at one site (Haith's laboratory in

Denver) and older babies at several other sites. In addition, younger babies are usually tested while lying on their backs, whereas older babies sit either in an infant seat or, more commonly, on a parent's lap. While these differences may have affected RT estimates, the fact that a saccade to a suddenly appearing peripheral target is a relatively simple response leads one to expect it to be more robust than %ANT.

In order to carry out a more complete developmental analysis, longitudinal data using the VExP at frequent intervals over a significant portion of infancy are needed. In addition, the data must be gathered using identical stimuli, methods, and procedures at all ages. Finally, because neither SDRT nor anticipation latencies are reported in the published studies, a new investigation will permit us to explore the nature of age changes and individual differences in these variables. In the following chapter, we describe the methods and procedures of a study that uses the guidelines stated above. We then undertake a comprehensive description of age changes for individuals and the group.

IV. DESCRIPTION OF THE STUDY AND METHOD

The primary focus of the present study is on providing an adequate database from which to investigate age changes and individual differences in infant saccades, both stimulus elicited (reactions) and expectancy driven (anticipations). Describing age changes for individuals and stability/instability of individual differences necessitates a longitudinal design. Thus, we followed 13 infants from 2 to 12 months of age. Assessments occurred each month from 2 to 9 months and again at 12 months. Ages in months represent chronological age since birth. For each assessment, infants visited the laboratory within 7 days of their birth date for the respective month. For all assessments, infants viewed the same 120-picture sequence of animated cartoon-like images that appeared to the left and right of visual center. Eye movements were recorded onto videotape and coded off-line by trained coders. Although the number of infants is not large, the number of behavioral observations is quite large. The primary data for the investigation consist of 11,229 trials (including reactive saccades, anticipatory saccades, and off-task trials) from 106 eye-movement recording sessions.

PARTICIPANTS

Parents who announced their infant's birth in the local newspaper were contacted by letter when their baby was approximately 6 weeks old and invited to participate in a longitudinal study of infant perception and cognition. The letters were followed by a telephone call during which the experimenter discussed the nature and purposes of the study. Eighteen families agreed to participate. Of the 18 infants (10 male, eight female) originally enrolled in the study, the parents of five males did not keep at least half their scheduled monthly appointments, and these infants' data were dropped from the analyses. For the 13 infants who constituted the final sample, we obtained usable data for 106 of the 117 possible visits. The 11 missing visits included two at 6 months (both missed their appointments), three at 7 months (two infants

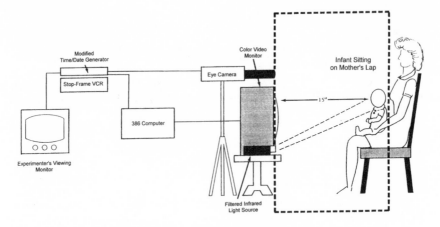

FIGURE 3.—Schematic of experimental apparatus (for details, see the text)

missed appointments, and one provided unusable data), three at 9 months (two infants left the study after the 8-month visit, and one family was unable to keep its appointment), and three at 12 months (two infants had previously left the study, and one family was unable to keep its appointment).

The infants' mean birth weight was 3,530 grams (SD = 556). Eight of the infants were born at term, two were born between 1 and 4 weeks early, and three were born between 1 and 4 weeks late. None of the infants experienced significant birth complications or chronic health problems.

Most of the infants were from white two-parent families. Infants' mothers reported their work status as homemaker ($N = 8$), clerical ($N = 2$), technical ($N = 2$), or professional ($N = 1$); fathers reported their work status as technical ($N = 4$), professional ($N = 4$), self-employed ($N = 2$), clerical ($N = 1$), student ($N = 1$), or unemployed ($N = 1$). The majority of the parents were college educated (mean years of education = 14.5 and 15.3, SD = 2.0 and 3.4, for mothers and fathers, respectively) and in their early thirties (mean age at infant's birth = 30.0 and 33.8, SD = 4.8 and 5.1, for mothers and fathers, respectively).

APPARATUS

Figure 3 shows the design of the apparatus. Visual stimuli were generated using the GRASP graphics animation software (Bridges, 1990) installed on a 386 Northgate PC-compatible computer and presented on a 48.25 centimeter SVGA NEC (Multisync XL) color video monitor. Infants were seated on one of their parents' lap in an opaque three-sided chamber measuring 1.22 meters wide by 1.22 meters deep by 1.67 meters high. The inside of the chamber was painted with an infrared-light-absorbent paint (Moore black, no. 215 80).

The infants' faces were illuminated by a Bausch and Lomb illuminator (catalog no. 31-32-42) fitted with filters restricting its emissions to a narrow band of wavelengths in the invisible, near-infrared range. Eye movements were recorded using a Cohu (4810) camera fitted with a chip sensitive to the same narrow band of invisible wavelengths produced by the filtered light source.[5] An image of the infant's eyes together with a 1/100-second time code generated by a Panasonic (WJ810) time/date generator and a digit indicating stimulus location and duration were recorded onto VHS format video-tape by a Panasonic (AG7300) video deck.

Because our time/date generator allows for 10-ms accuracy but the sampling rate for a VHS VCR is limited to 30 frames/second (33 ms/frame), we assigned each saccade latency to the nearest 33-ms bin. Therefore, when we discuss our data for individual saccades and theoretical performance levels (e.g., minimum RT), we limit ourselves to statements that assume no greater temporal resolution than that of our bins (e.g., "$X\%$ of our data fell into the 167-ms bin"). Note that we did not bin individual or group averages; we calculated standard arithmetic means from the binned individual trial data.

Stimuli

Stimuli consisted of seven computer-generated images that were randomly presented from trial to trial. Six images were animated (talking face, rotating head, opening/closing hand, running puppy, jumping stick figure, rotating wheel), and one was a static multicolored rectangle. The stimuli subtended approximately 7° of visual angle and appeared approximately 22° apart. Each stimulus was displayed for 700 ms against a black background, which remained black during the 1-second interstimulus interval. During the 3.4-minute period of data collection, the testing chamber was lit only by the stimulus display monitor and by a small amount of ambient light from the experimenter's viewing monitor, which entered the viewing chamber from behind the infant.

A total of 120 pictures was shown. The initial set of 20 pictures appeared in an irregular (IR), or unpredictable, sequence; the second set (40 pictures) appeared in an alternating (L-R) sequence; and the final set (60 pictures) appeared in an asymmetrical (L-L-R) sequence. Figure 4 shows a schematic example of the spatial and temporal arrangement of the three stimulus sequences. We hoped that, by using the three sequence types, infants would

[5] Specifically, the two-filter system uses a Corning 7-69 glass filter, to cut off the heat-producing wavelengths beyond 1,140 mμ, and a Kodak-Wratten 87c, which eliminates the shorter visible wavelengths.

VExP Sequence Types

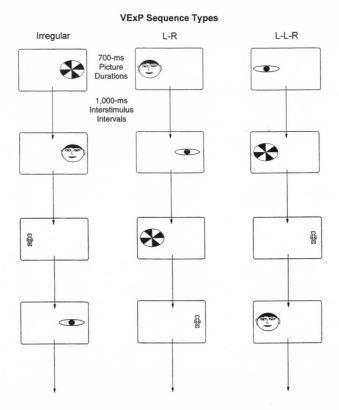

FIGURE 4.—Schematic examples of four VExP trials for each of the three sequence types shown in the present study. The sequence types are irregular (IR), left-right (L-R), and left-left-right (L-L-R). The rectangles represent the monitor at different points during the sequence, and the arrows indicate the sequential order of picture presentation. The pictures within the rectangles represent some of the stimuli that were actually presented and their location on the monitor. All pictures were presented for 700 ms and were followed by 1,000-ms interstimulus intervals. Picture location in the irregular sequence was randomly determined with equal numbers of pictures on the left and right. Pictures in the L-R sequence were presented in a spatially alternating pattern, and pictures in the L-L-R sequence were presented with a right picture occurring after every second left.

remain interested in viewing the pictures for more trials than if we used only a single alternating spatiotemporal pattern. We reasoned that sequential complexity would be increasingly important as infants got older and were no longer interested in simple alternation. In addition, the different sequence types allowed us to observe a wider range of responses than would have been possible had we used only an alternating sequence. For example, the IR and L-L-R sequences include situations in which the infant can make wrong-direction saccades during the anticipation interval that precedes a second same-side picture. As we show later, these wrong-direction saccades offer important converging evidence for determining minimum RT and its relation to age. Finally, we expected that some of the age differences in anticipation would be revealed in age by sequence type interactions.

DATA SCORING

Saccade latencies were coded off-line from the videotapes by human observers using a combination of slow-motion and stop-frame analysis. The coders' initial task was to learn the typical location where the infant's eye appeared on the playback monitor when he or she was fixating the left and right stimuli. Although the precise location of the eye on the playback monitor differed for different infants and depended on where the head was located when a particular trial began, coders had little difficulty reaching high levels of reliability in identifying the major fixation shift to the alternate side of the stimulus display screen. In making this judgment, it was necessary for coders to distinguish between eye movements to a stimulus location and those that were unrelated to the stimulus series. Furthermore, at this stage, the coders recorded trials for which off-task behaviors and equipment problems (such as the camera being out of focus) eliminated the possibility of gathering saccade latency data.

Coders classified any saccade away from the location of the previous picture and toward the other side of the visual field as a "shift." The magnitude of the saccade, its direction of movement, and its approximate final location were all considered in the coding of a saccade as a shift. Furthermore, a distinct movement of the eye was required for a movement to be classified as a shift. Head movements were not considered in our analysis.

In the case when an infant makes multiple shifts in relation to a given picture, only the latency of the initial saccade is coded at this stage of the data reduction. Any additional saccades beyond the first are considered "secondary" or "corrective" saccades, as they served to move the infant's visual regard further toward the location of the alternate-side stimulus. In this case, the initial saccade is termed *hypometric* because it falls short of its assumed target. If the infant made an initial hypometric saccade toward the alternate-side stimulus location but a secondary saccade landed a significant degree up or down or far overshot the location where the infant typically fixated the stimulus on that side, the trial was considered to be contaminated by off-task behavior, and no latency data were recorded. Furthermore, tiny microsaccadic adjustments involving barely perceptible movements of the infant's eye were not coded.

Reliability of coding was assessed by examining data from six infants (622 trials) that were coded independently by two observers. Reliability in correctly identifying the primary saccade, off-task trials, and missing data trials was very strong. Interrater agreement (percentage of trials on which coders agreed) was .96, and Cohen's kappa, controlling for chance agreement between coders, was .95. Coders were also highly reliable in determining, within one video frame (33 ms), the latency of the dominant fixation shift. Percentage agreement was maintained above 95% throughout the period of data coding. These high reliabilities were augmented when coding the remainder of the data by requiring multiple independent codings of especially problematic sessions and by collaborative resolution of any disagreements that remained.

We classified shifts as reactions or anticipations on the basis of a latency criterion. Determining the best estimate of this cutoff is one of the main goals of this *Monograph* and will be discussed further in Chapter V. Although reactions are always directionally appropriate, an anticipation can be classified as either a "correct" or an "error" anticipation. In the IR and L-L-R sequences, there exists the potential to make an inappropriate, or "error," anticipation between two consecutive same-side pictures, for example, shifting to the right after offset of a first left picture, only to have a second picture appear back on the left. Note that, because the infant is already looking to the left, the only codable shift that can be made is to the right. A "correct" anticipation, on the other hand, is a saccade to the location where the next stimulus will appear that has a latency shorter than the minimum possible RT. Except when specifically noted, when we refer to *anticipation rates* and *anticipation latencies*, we combine correct and incorrect anticipations.

An "off-task" code was given for trials in which the infant was judged to be distracted or inattentive to the stimulus display. Typical off-task behaviors included crying, turning around, yawning, or covering the eyes with a hand or an arm. Trials in which any of these behaviors prevented the infant from executing a distinct saccade to a stimulus location were coded as off task. Furthermore, the off-task code was given for excessively long reactions (greater than 1,000 ms), which we assumed to reflect generalized inattention due to boredom or sleepiness.

DESIGN

Our first concern was whether infants throughout the 2–12-month age range could be tested using the same stimuli, stimulus timing parameters, apparatus, and procedure. Our pilot work indicated that, by 2 months of age, most infants have sufficient physical control over their heads to allow them to view pictures while sitting on their parent's lap if their heads are resting on the parent's chest. We also observed several 12-month-old infants and confirmed that they found the animated pictures sufficiently interesting to watch for several minutes.

A major concern regarding the interpretation of our previous summary analysis was the problem of confounding between the implementation of the paradigm and the age of the infant. Through pilot work, one would expect that individual research labs would have chosen methods that work acceptably with the age group they tested. In addition, because older babies and younger babies have tended to be studied at different sites, age differences reported for infants across research sites may be unduly influenced by method variance. Of course, the nature of the paradigm and of infants may lead to a situation in which it is necessary to make constant changes in the task to ensure that it is appropriate for infants of various ages—as when in the visual preference paradigm one increases the complexity of the stimuli and shortens exposure times for older infants (Fagan, Singer, Montie, & Shepard, 1986). While it may have been possible to optimize infants' interest and/or comfort by altering the stimuli or procedures for different age groups, we believed that the benefits of using a uniform implementation of the paradigm for all ages outweighed the costs that would be associated with a confounding of age and procedure. Thus, for every assessment at all ages, babies viewed the same continuous stream of pictures that appeared on either side of visual center.

PROCEDURE

Parents brought their infants into the laboratory, where the experimenter explained the procedure and obtained informed consent. When the infant was judged to be in a quiet or an active alert state, he or she was placed on one parent's lap in a chair located in the experimental chamber (see Figure 3 above). Infants were positioned so that their eyes were approximately 60 centimeters from the center of the video monitor. The parents were cautioned that, when looking at the pictures themselves, they must remain still and not move their head or body.[6]

While the baby was being positioned, the monitor displayed large, high-contrast, black-and-white stimuli in each quadrant of the screen. These stimuli served to orient the infant's eyes to the video screen and allow the experimenter to focus the camera. When the experimenter had the infant's eyes in focus, the VHS deck was started, the room lights were extinguished, and the stimulus sequence was initiated. Filming continued until the infant became disinterested or fretful or until the full 120-picture sequence was completed. After data collection on the first visit, demographic information was

---

[6] It could be argued that we should have requested that parents wear opaque glasses during the procedure, but their ability to influence the behavior of their infants was limited. Our main concern was that they inhibit any of their own unconscious orienting movements, which they appeared to do.

31

TABLE 2

QUALITY OF THE DATA: NUMBERS OF SUBJECTS AND INFANTS' GENERAL ATTENTIVENESS BY AGE

Month	N	Mean (SD) Age in Days	Mean (SD) Duration of Testing Session per Infant in Minutes	Mean (SD) Percentage of Pictures Infant Was Off Task	Mean (SD) Number of Pictures Infant Was Actively Looking
2	13	62.2 (3.8)	2.71 (.49)	20 (16)	60 (17)
3	13	91.8 (3.2)	2.78 (.65)	29 (16)	55 (20)
4	13	124.5 (4.5)	2.76 (.45)	40 (18)	46 (19)
5	13	154.8 (4.0)	3.03 (.32)	36 (18)	54 (14)
6	11	187.7 (4.8)	2.83 (.63)	31 (16)	57 (23)
7	11	219.0 (3.5)	3.05 (.64)	31 (24)	61 (25)
8	13	252.0 (5.0)	3.15 (.49)	18 (12)	72 (19)
9	10	282.2 (6.6)	2.66 (.91)	19 (21)	63 (32)
12	10	369.4 (5.1)	2.46 (.91)	31 (18)	51 (26)

gathered from the parents. In addition, at every visit to the laboratory, parents were asked about their infant's behavior that day, including any illnesses from which the infant may be suffering as well as his or her recent eating and sleeping schedule.

General Evaluation of Data Quality

Since no previous VExP study has included infants across such a wide age range, we made some initial checks to determine the degree to which the quality of the data appeared comparable across months. Table 2 contains descriptive statistics for the numbers and ages of participants at each visit, the duration of the testing session, the percentage of pictures for which infants were off task, and the number of pictures at which infants actively looked (i.e., the number of trials that contributed latency data). The total looking duration measure shows that not all the infants remained interested in the stimuli for the entire 3.4-minute session. In fact, only at the ages of 7 and 8 months did a majority of infants continue to look for the entire 120-picture sequence.

Infant Attentiveness

The degree to which infants did attend to the pictures is shown by the indices of percentage off task and number of pictures actively fixated. Percentage off task is calculated using the equation

percentage off task = (number of off-task trials/total number of trials) × 100,

where the total number of trials excludes any trials with equipment problems or experimenter filming errors. This variable provides a measure of the quality of infant attention during the period of time before filming was stopped.

Means ranged from a low of 18% off task at 8 months to a high of 40% off task at 4 months. Differences between percentage off task at different months were tested using matched-sample *t* tests. Of all possible pairwise comparisons ($N = 36$), 10 were found to be significant at $p < .05$. However, because it is unclear what to expect about the relation between age and attentiveness, these tests were not guided by specific hypotheses. Because they are less active, younger infants may show less off-task behavior and longer looking times. Alternatively, older infants are likely to have longer attention spans and may therefore look longer. Since no predictions were made regarding age differences between groups and all possible pairs of means were tested, a Bonferroni correction for experiment-wise error is typically suggested (e.g., Keppel, 1982). With this correction, the alpha needed to reject the null is reduced to .001 (i.e., .05/36), leaving no significant age group differences. This conclusion is supported by observing that there are no systematic age trends in mean percentage off task (see Table 2).

One limitation to the percentage-off-task measure is that it may control for precisely the type of variability that is most revealing of age differences. Thus, we also examined the mean number of pictures actively fixated and found that it varied from a low of 46 pictures at 4 months (due partly to the high level of off-task behavior) to a high of 72 pictures at 8 months (due to a combination of long total looking times and low levels of off-task behavior). Similar to the percentage-off-task measure, no matched-sample *t* tests for pairwise comparisons between means remained significant after a Bonferroni correction. Considering both indices, we conclude that infants' general attentiveness in the task shows some variation from month to month but that this variation does not appear to be systematically related to age.

Attentiveness in Relation to State

Before beginning data collection at each visit, we made a global assessment to ensure that infants were in either a quiet or an active alert state. However, it is possible—especially in the early months—that infants' general

33

attentiveness may depend partly on the degree to which they are either tired or hungry at the time of testing. Therefore, at each visit, we asked the parent how long it had been since the baby had eaten and to estimate how long it would be until the baby's next nap. To determine whether the baby's general attentiveness was influenced by his or her level of hunger or fatigue at the time of testing, we used regression to predict the number of pictures actively fixated from parents' reports of time since last feeding and time until next nap.[7] In only one of 18 tests were these measures significantly related: at 2 months, the time since the infant was last fed was negatively related to the number of pictures seen ($b = -.25$, $p = .013$). In concrete terms, this indicates that, as the time since the last feeding increased by 40 minutes, 2-month-old infants saw an average of 10 fewer pictures.

The meaning of this one significant finding is dubious, and, as with the analysis of age differences in number of pictures seen, no significance remains following a correction for multiple tests. At no other age was either nap time or feeding time significantly related to the number of pictures seen (all unadjusted p's $> .10$). We also ran regression models for predicting the percentage of pictures for which the infant was off task from our two schedule variables. For no age group was either time since last feeding or time until next nap a significant predictor of percentage off task (all unadjusted p's $> .10$).

Overall, we found that infants were attentive to the stimuli at all ages. No consistent age differences were found in measures of general attentiveness or in the relation between our schedule variables and general attentiveness. Thus, we find little evidence to suggest major qualitative differences between age groups in infants' orientation to the stimuli, allowing us more confident interpretations of age differences in the performance measures. The final data set consists of 106 infant visits and a total of 7,597 trials for which infants made a saccade.

Before it is possible to move forward to describe age changes in saccade latency and anticipation, we must first address the issue of how to categorize saccades. An assumption made by researchers using the VExP is that the saccades that infants make when viewing the continuous presentation of peripheral stimuli are of two distinct types. In one category are the reactive saccades in response to the appearance of the peripheral stimulus. In the other category are expectancy-driven (anticipatory) saccades initiated by an internal command to move the eye to the location of an expected event.

Past research using the VExP has relied on a latency criterion (typically 200 ms) for determining which saccades fall into each category. The use of a latency criterion is supported by research using adults who view targets that

---

[7] Later, we used these "infant schedule" variables to inform our interpretations of age-related change in the performance measures but found them unrevealing.

step left and right in a partially predictable manner (Findlay, 1981; Horrocks & Stark, 1964). Research with infants has not provided a similar justification for a latency criterion, nor has it addressed the issue of possible age changes in the minimum latency to make a visually guided saccade.

Thus, in the next chapter, we take up the issue of how to determine empirically the latency criterion or criteria that we will use to categorize saccades as either visually guided or expectancy driven. An accurate classification of saccade latencies is the cornerstone of all VExP research; and, considering the possibility that the minimum RT decreases with age, it will be impossible for us to describe longitudinal age trends confidently unless we can be sure that the data are classified accurately. Only after we have solved this problem will it be possible to move on to the task of describing age changes and individual differences in information processing.

V. IDENTIFYING MINIMUM REACTION TIME

The description of age changes in RT and anticipation requires that these behaviors be measured on a common scale across development. Because the VExP is a free-looking procedure, for any given stimulus, infants' eye movements may be either elicited or internally produced; that is, the infants may either react to or anticipate the stimulus. The problem from a measurement viewpoint is to discover when babies are reacting and when they are anticipating.

A standard way to approach this issue has been to select a minimum latency required for babies to initiate an eye movement toward a visual stimulus. When a latency falls below this minimum, one concludes that the eye movement was initiated by the infant rather than elicited by the stimulus. Given the dramatic decline in average saccade latency during the first 8 months described in Chapter II, one can easily imagine that part of development consists of an age-related decline in the minimum latency required to initiate a reactive saccade. Assuming that the minimum latency declines with age, a consequence of using the same minimum RT for older and younger infants would be to underestimate age differences in average RT while overestimating age differences in percentage anticipation. In this case, RT and %ANT would not be measured on a common scale across development, and, as a result, the description of age changes would lead to an inaccurate picture of development. If minimum RT changes appreciably during the first year of life, then the developmental functions that we reported in Figures 1 and 2 above are necessarily inaccurate because all the studies shared a common definition of minimum RT.

In this chapter, we first review the two previous approaches for selecting minimum RT, the first described by Haith et al. (1988) and the second described by Canfield and Smith (1996). Haith et al. adopted a rational approach to the question of minimum RT, while Canfield and Smith adopted a more empirical approach. We evaluate both approaches, noting the strengths and weaknesses of each, and argue that both are overly conservative regarding the fastest speed with which infants can react. We then present three empirical methods for identifying the minimum RT, using data from the sample described in this *Monograph*. The first method, inspection of latency histograms, applies Canfield and Smith's (1996) approach at multiple ages, allowing the question of developmental change in minimum RT to be addressed. A second method, analysis of anticipation error latencies, provides a lower bound for estimating minimum RT by estimating maximum error anticipation latency. The third method, analysis of corrective saccades, provides compelling evidence for selecting a minimum RT that is substantially shorter than the currently accepted 200-ms value. In the final section, we conclude that minimum RT is more appropriately established at 133 ms and that, surprisingly, minimum RT does not change significantly from 2 to 12 months of age.

EARLY METHODS OF SELECTING MINIMUM RT

A Rational Approach

In their study introducing the VExP, Haith et al. (1988) considered the problem of determining a minimum RT value for babies and remarked that their selected criterion "had no direct data base, because we know of no published data on RTs for eye movements in 3.5-month-old infants" (p. 472). Existing data from adults had indicated that, when peripheral visual stimuli are presented in a spatially and temporally unpredictable manner, average RTs are in the range of 180–200 ms (Becker, 1972; Saslow, 1967b). Haith et al., therefore, chose to classify shift latencies less than or equal to 200 ms as anticipatory and shift latencies greater than 200 ms as reactive (equivalent to greater than or equal to 233 ms in our system).[8] Assuming that infants' reac-

---

[8] The binning method used with the data reported in this *Monograph* (also used in Canfield & Smith, 1996) prevents exact latency comparisons between our data and those of other laboratories. This is because video recording and timing systems vary slightly from one laboratory to another. The details of these systems require some explanation before one knows how to compare statements about latency values across different studies. The first principle of our method is based on the fact that we cannot achieve greater accuracy than one video frame (33 ms). This led us to bin our data and ignore the illusion of greater than 33-ms timing accuracy indicated by our 10-ms-accurate time/date generator. Instead, we used that information to indicate which was the correct latency bin for a particular shift. As a result, the binning process assigned shifts of 190 ms, 200 ms, and 210 ms to the 200-ms latency bin. Likewise, 220 ms, 230 ms, and 240 ms shifts were assigned to the 233-ms bin. Binning affects the recorded latency value of only a minuscule portion of our data. For example, only four 210-ms latencies were recorded in our entire data set. As a consequence, when we set our maximum anticipation latency at 200 ms, our minimum RT occurs at the next highest latency bin (233 ms), not at 201 ms—the value we and others used in the past reports. Note that some video-based systems can achieve field-level timing accuracy (16.7 ms). Furthermore, electro-oculograms and related non-video-based methods have a much faster sampling rate. Note further that the frame rate of non-VHS systems such as PAL is only 25

37

tion time is slower than that of the average motivated adult, the 233-ms minimum RT seemed an optimistic estimate of the fastest speed with which a young infant could react. Furthermore, Haith et al. were primarily interested in finding evidence for the existence of truly anticipatory saccades, and they wanted to protect against misclassifying reactions as anticipations. They believed that the 200-ms anticipation cutoff would provide such protection.

Haith et al.'s selection also appeared to be confirmed by adult data. Haith et al. showed two adult participants the same stimulus sequence that their infants were shown. The participants were instructed to respond to each event as rapidly as possible. The fastest shift for these adults was 196 ms (Canfield & Haith, 1991; Haith et al., 1988). In addition, in the only previous studies of saccade latency in infants, any fixation shift that occurred up to and including 200 ms after stimulus onset was considered to be unrelated to the stimulus, that is, too fast to have been elicited by it (Aslin & Salapatek, 1975; Salapatek, Aslin, Simonson, & Pulos, 1980). These three lines of argument provided the rationale for Haith et al. to select all shifts less than or equal to 200 ms as anticipatory.

There are two primary reasons why these criteria may have actually *overestimated* infants' minimum RT. First, an important difference existed between the procedures used with infants in the VExP and the procedures used with adults in visual reaction–time experiments (e.g., Becker, 1972; Saslow, 1967a, 1967b). Specifically, in the studies of adults, the central fixation point was immediately followed by the peripheral stimulus, a condition commonly known as a *no-gap condition* (Hood & Atkinson, 1993). In the no-gap condition, average RT for adults is about 200 ms. However, saccade latency to the peripheral stimulus can be significantly reduced by introducing an interstimulus interval between the offset of the fixation point and the onset of the peripheral stimulus. This reduction has come to be known as the *gap effect.* In the gap condition, when the fixation point is extinguished at least 200 ms in advance of the onset of the peripheral stimulus, average adult RT is approximately 150 ms (Saslow, 1967a). This time difference between the no-gap and the gap conditions is widely thought to reflect the additional time needed to disengage covert attention from the central fixation point in the no-gap condition (Fischer & Weber, 1993; Johnson et al., 1991; Posner & Peterson, 1990). The gap may also allow for initial stages of saccade programming or for a more general increase in arousal and readiness to respond (Luce, 1986).

Importantly, implementations of the VExP use a form of the gap condi-

---

frames/second, or 40 ms/frame (e.g., Hood & Atkinson, 1993). These differences can be the source of some confusion when making the kinds of distinctions that we make in this *Monograph,* which is why we limit ourselves to considering raw latency values as different only if they are assigned to different bins.

tion, with gap durations ranging from 500 to 1,500 ms. Hood and Atkinson (1993) show that, as do adults, infants react significantly more quickly when there is a gap of at least 240 ms between stimuli. Thus, rather than compare infant VExP to adult no-gap conditions, it is more appropriate to compare infant VExP to adult gap conditions.

A second possible reason why minimum RT for infants may have been overestimated concerns the instructions given to the adults who viewed the same stimulus sequences seen by the infants. Namely, adults were instructed to *react* as quickly as possible after the stimulus appeared. It is likely that this instruction was interpreted by the participants as a command to avoid anticipations. Active avoidance of anticipatory responding may have required additional processing time, which may have slowed the adults' reactions.

An Empirical Approach

Canfield and Smith (1993, 1996) recognized the limitations of deriving a minimum RT for infants solely from adult data. In a study of number-based expectations in 5-month-old infants, they proposed an empirical method for determining minimum RT from visual inspection of infant latency data. They reasoned that the distribution of all saccadic latencies is composed of two overlapping distributions, one consisting of anticipations and the other of reactions. The distribution of reactive saccades should appear positively skewed, as is typical with response-time distributions (Luce, 1986). The distribution of anticipatory saccades, on the other hand, should appear more uniformly distributed because anticipatory responses are more self-selected than elicited. The overlap between the two distributions begins at some point after the onset of the picture (a saccade latency of 0 ms). At the upper end of the overlap there should be a sharp discontinuity, corresponding to the point at which the reactive portion of the distribution becomes predominant. By noting the latency bin at which the discontinuity is located, it is possible to choose a minimum RT that is derived directly from the data rather than being inferred from adult performance in a similar task.

In both studies of 5-month-old infants, Canfield and Smith (1996) found a marked discontinuity in the saccade latency distribution; the discontinuity occurred 200 ms after picture onset (see Figure 5). Furthermore, the discontinuity was located at the same latency value for both the predictable and the unpredictable experimental conditions and was likewise reflected in individual data. Canfield and Smith used this empirical criterion to define the minimum RT as 200 ms in their study, with any shift occurring *more quickly* than 200 ms (i.e., falling in the 167-ms or a lower bin) being defined as an anticipation. This finding contrasts with previous VExP research, all of which classifies 200-ms latencies as anticipations. As shown in Figure 5, nearly 3% of all laten-

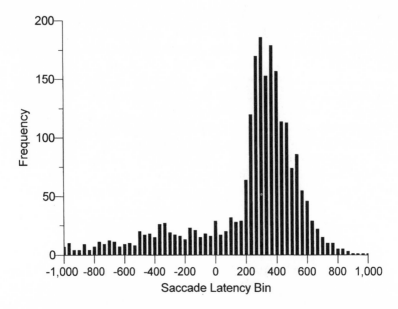

FIGURE 5.—Frequency histogram of saccade latencies (adapted from Canfield & Smith, 1996, experiment 1; N = 2,168 shifts). Latency bin widths are 33.33 ms (one video frame). There is a marked discontinuity in the distribution beginning at the 200-ms latency bin. This discontinuity was located at the same point for both the IR and the L-L-R stimulus conditions and was likewise reflected in the individual subject data. Canfield and Smith used these data to recommend a minimum RT of 200 ms. Used with permission from Canfield and Smith (1996).

cies fall in the 200-ms bin. Looked at another way, had we included these shifts in the anticipation distribution, they would have accounted for more than 11% of all the anticipatory shifts, thus representing a potentially serious inflation of our anticipation data.

The Canfield and Smith findings represent the first report using infant latency data to determine minimum RT. By concluding that the minimum was 167 ms rather than the "greater-than-200-ms" value used in previous studies, Canfield and Smith's results might be seen as posing a challenge to previously published research. However, issues remain to be resolved before such a conclusion can be supported. For example, most VExP research has been conducted with 2–3-month-old infants, but Canfield and Smith's participants were 5 months old. Therefore, finding a shorter minimum RT may simply indicate that minimum RT declines with age. It is possible that younger babies require more time to generate a saccade—not just on average but as a minimum.

In addition, the methods used by Canfield and Smith can be criticized for several reasons. First, the objectivity of any analysis that is based solely on visual inspection and judgment about what is a "large" difference between adjacent frequency bars can be questioned, and, although a discontinuity in their data was unmistakable to the eye, no formal analysis of the point of separation between the two distributions was undertaken. Second, it is possible that the discontinuity in the distribution does not correspond to the best estimate of minimum RT. Recall that the discontinuity occurs when the magnitude of the reaction distribution surpasses the magnitude of the anticipation distribution. The minimum RT, however, may occur at a point when the anticipation distribution is predominant, thereby obscuring the beginning of the reaction distribution. Third, although the existence of independent distributions is often used as prima facie evidence for the existence of distinct underlying processes, such a conclusion is unwarranted without additional converging evidence (Howe, Rabinowitz, & Grant, 1993).

PRESENT METHODS OF SELECTING MINIMUM RT

Despite the problems with the Canfield and Smith method, its primary strength is its simplicity. It is instructive, therefore, to apply this method to our longitudinal sample of infants. As a first approximation, this straightforward method will provide clues concerning the possibility of developmental change in minimum RT. It will also serve as an introduction to two additional methods that allow more precise specification of minimum RT.

Inspection of the Latency Distributions

The most straightforward way to find where the anticipation and reaction distributions meet is to construct histograms of saccade latencies for individuals and to note the latency value corresponding to the trough between modes, if one exists. This method was used by Horrocks and Stark (1964) and by Findlay (1981) in their studies of reaction and anticipation in adults, but they had the luxury of having many hundreds of latencies for each of only four (or fewer) participants. In the present study (as in Canfield & Smith, 1996), we have an average of only about 70 saccades for each participant, but we have 106 individual distributions (13 participants × 9 ages − missing data). General patterns of age change can be more readily discerned when the latencies for all participants in a given age group are aggregated, resulting in nine distributions, each having 500–900 latencies.

The nine percentage histograms corresponding to the nine age groups

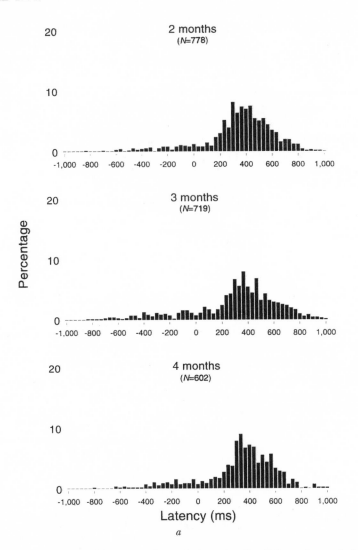

FIGURE 6.—Percentage histograms of saccade latencies plotted for each month. Latency bin widths are 33.33 ms (one video frame). In accord with Canfield and Smith's (1996) findings, a marked discontinuity can be seen at the 200-ms bin for 5-month-olds. Beginning at 6 months, the discontinuity occurs at the 167-ms bin.

are shown in Figure 6. In each panel, all latencies are plotted for a given month across all infants and all three sequence types. The bar width is equal to the temporal resolution of our video-based coding system (33 ms). Specifically, the height of each bar represents the percentage of latencies occurring during that 33-ms sampling period. The length of the abscissa repre-

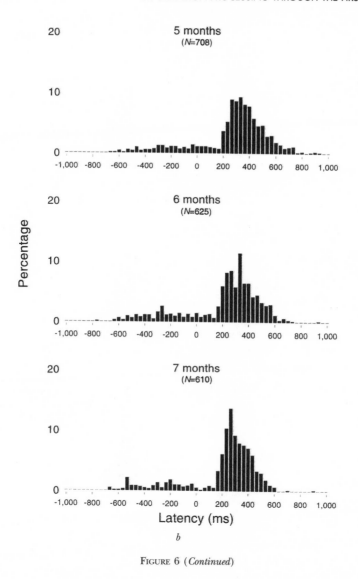

FIGURE 6 (*Continued*)

sents a 2-sec period of time during which a saccade could occur in relation to a given stimulus. This time period begins with the offset of the previous stimulus (−1,000 ms) and ends 300 ms after the offset of the stimulus under consideration for that trial (+1,000 ms). For every stimulus, only one latency value was given.

Several features of these distributions can be noted from a visual inspection. First, the data for the 5-month-old infants show a discontinuity at the

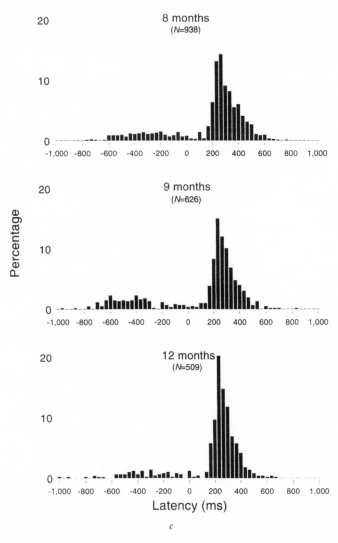

FIGURE 6 (*Continued*)

200-ms bin, replicating the finding from Canfield and Smith (1996). And, with respect to age differences, the presence of the discontinuity appears to change systematically with age.

The distributions at 2–4 months show no discontinuity consistent with a sudden dramatic rise in the frequency of fixation shifts when reaction becomes possible, although the 2-month-olds show a moderate increase from the 133-ms to the 167-ms bins. The relatively smooth increase in the percentage of shifts from the time of stimulus onset to the maximum of the distribution prevents us from using this method to determine minimum RT for infants younger than 5 months of age.

Inspection of the distributions from the later assessments reveals a discontinuity. For the 5-month-olds' distribution, there is a subtle decline in the percentage of shifts occurring from −33 ms up to the 167-ms bin after stimulus onset. At 200 ms and later, however, the percentage of shifts per latency bin increases dramatically and does not decrease until after the maximum of the distribution. Beginning at 6 months, the discontinuity occurs earlier. For all distributions from 6 to 12 months, there is evidence of a rapid increase in the response rate beginning at the 167-ms latency bin. This pattern of data suggests that minimum RT declines from 167 ms to 133 ms sometime between 5 and 6 months. Unfortunately, we have no clear indication from this method what the value is for younger babies. Because the visual inspection method did not work for infants younger than 5 months, and because visual inspection of the histograms can provide only an upper bound to the range of latencies containing the minimum RT, additional methods of investigation are necessary before an appropriate minimum RT can be chosen.

Anticipation Error Latency

Having estimated an upper bound for minimum RT through visual inspection of the latency histograms, it is also possible to set a lower bound by determining the maximum latency at which anticipation errors occur. Anticipation errors are anticipatory shifts that are followed by the appearance of the stimulus at the previously abandoned location. They occur in the context of the IR and L-L-R sequences, when babies are confronted with a situation in which two or more pictures appear sequentially in the same location. If reactive and anticipatory saccades represent somewhat distinct categories of eye movements, then the maximum latency at which infants make anticipation errors may help determine the minimum RT. The process of programming an anticipation will be canceled if new visual stimulation appears at the currently fixated location (Aslin & Shea, 1987; Stark, Michael, & Zuber, 1969). However, there is a point beyond which the saccade program cannot be canceled.[9] The maximum anticipation latency, the latest time at which infants make anticipation errors, represents this point of no return.

[9] This situation, in which new stimulus information affects ongoing saccade programming, is similar in many respects to variants of the "Wheeless paradigm" (Wheeless, Boynton, & Cohen, 1966, cited in Aslin & Shea, 1987). This paradigm has been studied and refined extensively, resulting in the development of the "pulse-step," "pulse-overshoot," and "pulse-undershoot" variants. Our situation differs from these in two possibly very important respects: (1) Rather than the stimulus information interfering with reactive saccade programming, it interferes with anticipatory saccade programming. (2) In the reactive case, it

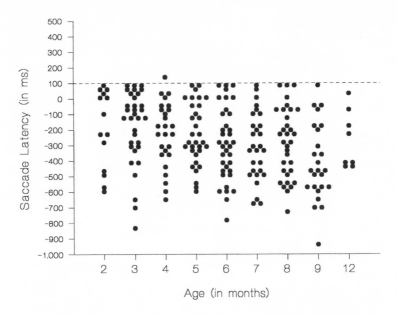

Figure 7.—Latency of anticipation errors as a function of age for all infants ($N = 235$, 2–12 months). Each point represents the latency of a single anticipation error. A latency of 0 ms corresponds to picture onset. The dashed line marks 100 ms after picture onset.

Figure 7 shows the distribution of anticipation errors for all infants for each age group. Each symbol in the plot represents one error shift falling in a particular latency bin. Of the 235 fixation shifts, with the exception of one saccade by one infant at the age of 4 months, no anticipation errors occurred later than the 100-ms bin for any infant. Thus, the infants in our study appear to have a maximum anticipation error latency of 100 ms. Interestingly, this maximum did not change as a function of age. Together with our inspection of latency histograms, this analysis indicates that minimum RT is between 100 and 167 ms after picture onset.

The question of developmental change in minimum RT remains open. Inspection of latency histograms indicated a possible drop in minimum RT between 5 and 6 months of age. Analyses of anticipation error latencies suggest, however, that minimum RT may not change over infants' first year. In the next section, we conduct an analysis of corrective saccades that will help identify an appropriate minimum RT and also determine whether our estimate should be adjusted as infants get older.

---

would represent a "pulse-return" variant of the paradigm. We are not aware of any studies with adults or infants that have explored saccade programming in these two situations.

Corrective Saccades

Researchers investigating adult saccadic eye movements have developed several methods for estimating the point of separation between anticipatory and reactive saccades. These studies have typically focused on questions of whether one can find evidence for a distinction between reactive and anticipatory saccades by studying their latencies, velocities, amplitudes, and main sequences. In an early study, Young and Stark (1962a) studied saccades in adults who tracked a stimulus that abruptly stepped from left to right every 500 ms. Individuals tracked the stimulus for a few cycles and then began to anticipate the movements on a substantial proportion of trials. Saccades that began before the stimulus had stepped were less accurate than those that began 100–130 ms after the step. Young and Stark interpreted the difference in accuracy as reflecting the fact that memory, rather than visual information, was guiding the saccades with latencies shorter than about 100 ms; that is, these saccades were anticipatory.

A more recent approach to this issue, which involves carefully analyzing saccade metrics, lends additional support to the conclusions of Young and Stark. Bronstein and Kennard (1986) studied the amplitude–peak velocity relation for anticipatory and reactive saccades to help distinguish the two saccade types. They found that saccades shorter than 100 ms had slower velocities than the longer-latency saccades, indicating the possibility of a different source for the saccade pulse. In addition, they found that the saccades with lower velocities tended to fall short of their target—a finding consistent with prior research by Findlay (1981).

These three studies converged on two common conclusions. First, anticipatory saccades are more hypometric than are reactions; that is, they are more likely to undershoot the target. Both Bronstein and Kennard (1986) and Findlay (1981) found that reactive saccades traveled an average of 90% of the distance to the target but that the average anticipatory saccade traveled only 75% of the distance. Second, minimum RT for adults making saccades to somewhat predictable targets is about 100 ms.

The findings from studies of adults point to the possibility of using objective measures of saccade accuracy to categorize saccade latencies as either reactions or anticipations. Unfortunately, these methods are difficult and very expensive to apply to an infant population. One of the strengths of the VExP is that it is not necessary to track the exact location of the infant's fixation. Only the direction (left or right) and approximate amplitude of the saccade are required in order to determine reliably whether the infant shifts fixation to the stimulus location (Canfield et al., 1995). This feature of the paradigm allows infants to view stimulus displays without researchers having to attach sensors to the infants' faces or rigidly constrain the infants' head movement as required for the calibration of video-based tracking systems (Aslin, 1985;

but see Bronson, 1994). Finally, although these more precise methods can be used with infants younger than 4 months, they become increasingly difficult to employ as the infants gain motoric competence. However, we can apply the basic logic of these methods to our infant data.

Because anticipatory saccades are quite hypometric, a relatively large secondary saccade, known as a *corrective saccade*, is often required if an individual is to orient accurately to a peripheral stimulus. In the Findlay (1981) study, anticipatory primary saccades were followed by corrective saccades 50%–70% of the time, while Bronstein and Kennard (1986) reported corrections for anticipations in an average of 55% of cases. By contrast, reactive saccades were followed by corrective shifts on only 2%–18% of the trials in the Findlay study and on an average of 8% of the trials in the Bronstein and Kennard study. These findings mirror those of Young and Stark (1962a, 1962b; see also Stark, 1971) by confirming that anticipatory saccades are often inaccurate.

Although the eye-movement-recording methodology used in the present study does not allow us to measure precise fixation location, we can easily detect when a primary saccade is followed by a corrective saccade. This allows us to use the proportion of corrective saccades to classify the primary saccades as either anticipatory or reactive. Specifically, we should observe a lawful relation between primary saccade latency and the incidence of corrective saccades, as Bronstein and Kennard (1986) and Findlay (1981) did in their studies of adults. Anticipations should be significantly more likely to be followed by a corrective saccade, and, if there is a latency bin at which the frequency of corrective saccades abruptly declines, that would correspond to the minimum RT.

For the purpose of documenting corrective saccades in our data, we selected only those saccades with latencies between −100 and +267 ms. This region spans three types of latencies: (1) those that must be anticipatory because they precede the stimulus; (2) those that are almost certainly reactive since they follow the stimulus by an amount greater than that suggested by our inspection of latency distributions; and (3) those that cannot be easily classified solely on the basis of their latency characteristics.

We defined a corrective saccade as an eye shift toward the stimulus that followed a primary saccade. Secondary shifts excluded from the analysis included those that were closely followed by some indication that the infant was distracted (e.g., looking down, turning around) and those that were directed away from the currently displayed stimulus. Nearly all corrective saccades occurred within 200–300 ms of the primary saccade, and very few corrective saccades occurred after the stimulus disappeared. Therefore, on the basis of our current estimate of the upper bound for minimum RT, we reasoned that shifts occurring as late as 200 ms after offset were likely to be a search for what disappeared rather than a late attempt to better fixate a stimulus long since removed from sight.

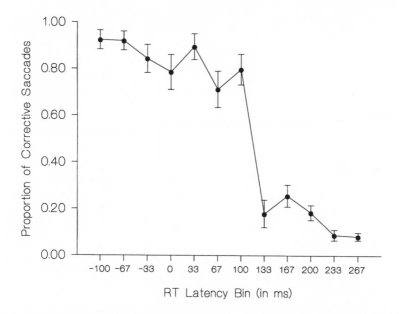

FIGURE 8.—Proportion of primary saccades followed by a corrective saccade as a function of RT latency bin for all subjects at all ages (2–12 months). Latency bins are 33.33 ms (one video frame). Error bars represent the standard error of the mean.

None of the infants made primary saccades in every latency bin from −100 ms to 267 ms at every age. Therefore, our data are somewhat sparse, particularly for the first few latency bins immediately following stimulus onset—the bins most relevant to determining a minimum RT. One possible reason that these bins include very few saccades is that, were the peripheral stimulus to appear during an early phase of anticipatory saccade programming, the program may be aborted, thereby allowing a more accurate visually guided saccade to be programmed and deployed (Aslin & Shea, 1987; Horrocks & Stark, 1964). To maximize the number of data points in our analysis, we will first present the data averaged across all participants and ages. Then we will consider whether the conclusions reached from that analysis must be qualified by patterns of findings for individual ages and infants.

Figure 8 shows a plot of proportion of primary saccades followed by a corrective saccade as a function of latency bin, with each point representing an average over infants and across the nine age levels. The plot shows quite dramatically that infants are substantially more likely to make a corrective saccade when the primary saccade either precedes the stimulus or follows it by a period shorter than 133 ms. Across all age groups, an average of 84% of the primary saccades occurring from −100 to 100 ms were followed by a corrective saccade, while only 16% of the primary saccades occurring from

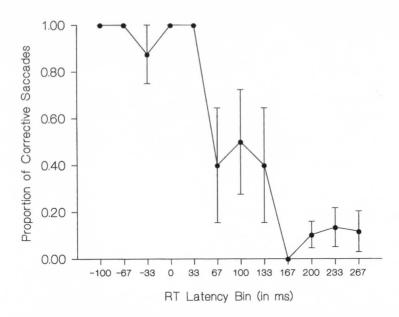

FIGURE 9.—Proportion of primary saccades followed by a corrective saccade as a function of latency bin for all subjects at 2 months. Latency bins are 33.33 ms (one video frame). Error bars represent the standard error of the mean.

133 to 267 ms were followed by a corrective saccade, proportions strikingly similar to those reported for adults.

We wanted to assess the significance of the difference in proportions of corrective saccades as a function of the latency bin value and especially to test for the significance of an age by bin value interaction. Therefore, we computed a repeated-measures ANOVA comparing the average proportion of corrective saccades for bins from −100 to 100 ms to the average proportion for bins from 133 to 267 ms. Results of the ANOVA revealed a significant main effect of bin value ($F[1, 590] = 696.74$, $p = .0001$) and age ($F[8, 590] = 2.68$, $p = .007$), but the age by bin value interaction was nonsignificant ($F[8, 590] = .97$, $p = .461$). This analysis provides no evidence that the pattern of differences in the frequencies of corrective saccades differs as a function of age.

Given the apparent age differences in the minimum RT revealed by our visual inspections of the latency distributions, we were somewhat surprised to find no age by bin value interaction for the proportion of corrective saccades. Therefore, we plotted graphs of corrective saccades by bin value for each age group. With the exception of the data from the 2-month assessment, every age group showed a pattern similar to that shown in Figure 8. The pattern was less clear for the 2-month data. As shown in Figure 9, the 2-month-olds showed a sizable decrease in corrective saccades beginning at 67 ms. The proportion remains at about 40%–50% until it drops to nearly 0% at the 167-ms latency bin.

We do not know what the cause might be of the first decline beginning at 67 ms, and we would want to see it replicated before we engaged in much speculation, but it is not the result of clearly different patterns of findings for different infants. Neither is the pattern consistent with the interpretation that younger infants make many multiple saccades; were it so, the rate of corrective saccades would have been high regardless of the bin value. Furthermore, unlike the reports of Aslin and Salapatek (1975) and Salapatek et al. (1980), our data did not lead to the conclusion that multiple saccades are pervasive during infancy, even for our youngest infants (see also Hainline, 1984). Instead, we find that some 2-month-olds make a high proportion of multiple saccades but that, by 3 or 4 months, they are almost nonexistent. We suspect that the lack of multiple saccades may be due to the large, bright, animated stimuli that we used. Given the centrality of multiple saccades to the general understanding of infant oculomotor competence, it is interesting that no studies have explored the effects of stimulus intensity or size on their prevalence.

Finally, we inspected plots for each infant at all ages and found no evidence to suggest individual differences in the location of the bin at which the proportion of corrective saccades declined. Thus, we conclude that a reasonable choice for the minimum RT is 133 ms. In addition, we conclude that this minimum does not appear to change much with age, at least throughout the 3–12-month age range in our sample. Given that our analysis of maximum anticipation error latency found no differences from 2 to 12 months, there is sufficient justification to select 133 ms as the minimum RT for 2-month-olds as well. The ambiguous findings from the analysis of corrective saccades may have been influenced by a small number of shifts in some bins, coupled with a few double shifts.

CONCLUSIONS

The general convergence of our findings from the analysis of latency histograms, anticipation errors, and corrective saccades leads us to conclude that the minimum RT for our babies—that is, the fastest a baby can use visual information to elicit a saccade in our implementation of the VExP—is 133 ms. This conclusion is supported by the following findings: (1) Saccade latency distributions for stimuli that are spatially and temporally predictable show a sharp discontinuity after 4 months of age, indicating distinct saccade types. (2) Anticipation errors are not made later than 100 ms after stimulus

51

onset. (3) The proportion of corrective saccades declines dramatically when the primary saccade latency is 133 ms or longer.

In addition, the data suggest that this minimum RT is uniform across infants and does not change between 2 and 12 months because there were no significant age differences revealed by the analysis of the proportion of corrective saccades by bin value. Furthermore, anticipation errors showed a maximum latency of 100 ms or shorter for all age groups and all infants, with the exception of one saccade for one 4-month-old infant. Finally, we found no evidence of consistent individual differences in the pattern of corrective saccades.

Our conclusions must be tempered by stating that minimum RT and maximum anticipation error latency are specific to particular experimental paradigms and that our value of 133 ms should therefore not be interpreted as an absolute cutoff that can be applied to any paradigm for measuring infant saccades. Nevertheless, we believe that our minimum will apply to a broad range of implementations of the VExP and that the currently accepted minimum RT is a potential source of inaccuracy in previously published research using the paradigm. (We revisit this issue in Chapter VIII.)

Having a minimum RT in hand, it is now possible for us to use a latency criterion to categorize saccades as being either elicited by the stimulus (reactive) or emitted by the infant (anticipatory). Further, we believe that this categorization need not be adjusted as a function of age. We are now in a position to describe developmental functions and individual differences in RT and anticipation for our sample of infants. In Chapter VI, we describe developmental functions and analyze individual differences in RT. Trial-to-trial variability in RT (SDRT), %ANT, and anticipation latency (ANTL) will be examined in Chapter VII. In Chapter VIII, we assess the practical effects of using the 133-ms minimum RT value on our description of age changes by comparing growth curves using the 133-ms value with the growth curves obtained using the 233-ms value, enabling us to gauge the degree to which existing research findings may be biased by past use of a different minimum RT value.

VI. DEVELOPMENTAL FUNCTIONS AND INDIVIDUAL DIFFERENCES IN REACTION TIME

As we noted above, a complete understanding of development rests on a careful description of intraindividual change over time as well as interindividual differences in the pattern of intraindividual change. In the present chapter, we carry out this descriptive task for the RT measure. Our analytic plan involves four steps. First, we carry out several general analyses designed to help determine which data are best suited for modeling individual growth and the stability of individual differences. Because noteworthy differences in the pattern of age-related change may be found for the three sequence types, separate growth-curve analyses may be necessary for each. Thus, in the second step, we look for the general presence, direction, and magnitude of age-related changes in RT at the group level, but we look separately for sequence IR, L-R, and L-L-R. These growth-curve analyses will enable us to describe in detail the pattern of change (i.e., the shape of the growth function) during the first year of life. In the third step, we pursue mathematical descriptions of individual growth, using both polynomial and inherently nonlinear models. Finally, in the fourth step, we consider individual differences in RT, first by noting the presence of differences in the shape of individual growth functions, then by analyzing the stability of individual differences in RT over the course of the first year.

CHOOSING A MEASURE OF CENTRAL TENDENCY AND IDENTIFYING OUTLIERS

Before commencing with the analysis of reaction time, it is necessary to address two questions of particular relevance to the study of RT distributions: how should we handle extreme individual data points that may be overly influential in the computation of averages, and what is the most appropriate measure of central tendency to use for the RT distribution?

53

Choosing a Measure of Central Tendency

In previously published VExP research, the sample median has typically been chosen as the measure of central tendency for RT. However, Miller (1988) reports that, for positively skewed distributions, which are common for RT distributions (see Figure 6 above), the sample median is not an unbiased statistic; in fact, it systematically overestimates the population median, sometimes by as much as 50 ms. Miller's simulations showed that this bias is especially problematic when medians are used to compare different experimental conditions, each having a different number of trials, as is the case in VExP research when comparing baseline to postbaseline performance. The sample median, in this case, will overestimate the population median by a greater amount in the condition having fewer trials. This may lead to the false conclusion that the baseline RTs are significantly slower when the difference may be due exclusively to differential overestimation of the population median.

Miller's observations have important ramifications for all research using RT as a dependent measure, but they are especially significant for VExP research because many of these studies have used designs that compare baseline performance with postbaseline performance (e.g., Canfield & Haith, 1991; Canfield et al., 1995; Haith et al., 1988; Jacobson et al., 1992). The baseline RT is typically calculated from the first 5–10 RTs, but the postbaseline RT is typically based on 50–75 RTs. Furthermore, given the inevitable existence of missing data in infancy research, it is not uncommon for the baseline RT to be calculated on fewer than five observations.

According to Miller's (1988, p. 541, table 1) simulations, assuming a modest mean-median difference of 54 ms in the population, if the baseline median RT is calculated on four observations and the postbaseline median RT is calculated on 50 observations, one would expect the baseline median to be 25 ms greater than the postbaseline median owing to differential overestimation bias alone! Given that the absolute magnitude of a statistically significant drop from baseline to postbaseline median RT is frequently only modest (averaging 50 ms in the studies cited above), one must be concerned that a finding of statistical significance based on a small difference between medians is erroneous and would not have been found using a statistic that gives an unbiased estimate of the central tendency of the population distribution. Miller's simulations suggest that the difference in population medians in the example above is only 25 ms—half what the sample medians indicated.

Miller recommends using the sample mean rather than the median, particularly when comparing small distributions with very different numbers of cases. The sample mean remains an unbiased estimator of the population mean in these situations, and the standard errors of the mean and median are comparable even when distributions are significantly skewed (Miller, 1988). Therefore, in our subsequent analyses that are based on a measure of central tendency, we use the mean of RT as our dependent measure.

Method of Dealing with Outliers

A common problem when dealing with distributions of response latencies is the presence of extreme scores. Very long latencies are typically believed to reflect the influence of additional psychological processes besides those that the study is designed to assess (e.g., a change in the state of arousal or a momentary distraction). A primary reason why researchers often use the median RT is the obvious fact that the sample mean is sensitive to extreme scores. Therefore, when using the mean, a method for eliminating outliers is recommended. Importantly, our outliers appeared to be randomly distributed across stimulus conditions (IR, L-R, and L-L-R). Thus, our concern with outliers was related not so much to the possibility of systematic bias as to the possibility that outliers would increase random error. In dealing with outliers, we used a combination of techniques suggested by Ratcliff (1979, 1993), who investigated the effects of different methods of dealing with reaction-time outliers on the power of the ANOVA to detect mean differences between groups.

Because our RTs came from a free-looking procedure, there was a natural cutoff value for long reactions. That is, whether or not the infant had responded, the next stimulus appeared at a predetermined time. Therefore, we allowed the longest RT to be 1,000 ms (300 ms after stimulus offset). Even the very young infants in our studies produced few RTs that were as long as 1,000 ms (see Figure 6 above). Typically, any RT longer than 1,000 ms was associated with off-task behaviors such as looking down, prolonged eye closure, or turning to reference the parent. Consequently, we converted the few RTs greater than 1,000 ms to off-task trials.

For some infants, outliers continued to be a concern for the remaining RTs between 133 and 1,000 ms. Using Ratcliff's simulations with ex-Gaussian distributions as a guide, we decided to use a 2-standard-deviation cutoff to eliminate the remaining outliers. Ratcliff found that this method allows for F tests with high power, and, when we used each infant's distribution of RTs from a given session as a reference, virtually all extreme scores fell more than 2 standard deviations from the infant's mean RT for that session. Interestingly, Ratcliff's simulations also revealed that using the median RT as a way to reduce the influence of outliers compromises the power of F tests (when testing for a 30-ms difference in μ). This provides yet another reason why the median RT should be avoided as a measure of central tendency.

GROWTH CURVES FOR RT

Following the recommendations of Burchinal and Appelbaum (1991) and Wohlwill (1973), we use a developmental function approach to describe age changes in RT. We begin by determining the presence and direction of developmental change in the group and then move to a more careful description of shape using linear regression methods. We then move beyond the analysis of the group data and fit individual polynomial growth curves. The polynomial growth models are evaluated in relation to the data and theory about the nature of growth in this domain, leading us to consider an inherently nonlinear function, the exponential, to describe the age-related change in RT.

Presence and Direction of Age-Related Change

Having chosen the sample mean as an appropriate measure of central tendency, and having decided that for a given infant within a session any RTs beyond 2 standard deviations of the mean will be considered outliers, we can proceed with the analysis of RT. It remains for us to decide whether to carry out separate analyses of growth for each of the three sequence types or to combine all the RTs for a given session without regard to which sequence they originated from. Our decision will depend on whether mean RT shows different trends as a function of age for each of the sequence types employed in the study. If indeed there are different trends for each sequence type, growth-curve analyses for each type will need to be conducted.

It is important to clarify at the outset of this analysis that our main purpose is to determine whether sequence type is likely to be relevant to our subsequent investigations of individual growth functions. This study was not designed to investigate in a thorough manner either the effects of sequential complexity or the possible interactions between complexity and age of assessment. We included the three sequence types in an attempt to observe infant oculomotor behavior in a range of stimulus conditions and to maintain infants' interest over an extended period of time. As a result, any conclusions that we draw about the role of sequence type will be fully confounded with the order in which the sequences occur and the number of pictures that the infant previously viewed. Conclusions about the effects of these variables will require studies with additional subgroups to allow for the counterbalancing of sequence type, sequence duration, and order of sequence occurrence. Given the magnitude of the resources required to carry out this type of research, our hope is that the present study will prove sufficient to indicate whether the future expenditure of such additional resources is warranted.

TABLE 3

MEAN (SD) REACTION TIME (in Milliseconds) AT EACH MONTH
FOR ENTIRE SESSION AND INDIVIDUAL SEQUENCE TYPES

		SEQUENCE TYPE		
AGE	GLOBAL[a]	IR	L-R	L-L-R
2 months	440	455	424	451
	(57)	(74)	(64)	(65)
3 months	439	428	437	443
	(69)	(103)	(60)	(99)
4 months	418	434	401	411
	(67)	(144)	(60)	(76)
5 months	382	433	371	383
	(58)	(122)	(53)	(71)
6 months	354	350	346	377
	(56)	(61)	(57)	(82)
7 months	333	349	320	335
	(50)	(46)	(60)	(57)
8 months	318	316	312	330
	(60)	(29)	(60)	(77)
9 months	287	290	280	296
	(42)	(43)	(44)	(54)
12 months	285	294	281	292
	(30)	(51)	(31)	(54)
Mean (SD)	362	372	352	369
	(61)	(66)	(59)	(59)

[a] *Global* denotes that the variable was calculated without regard to sequence type.

The Shape of the Developmental Function

Table 3 contains mean RT calculated both with and without regard to sequence type (global) and listed separately for each month. When averaged across infants for each month, mean RT in the L-R sequence is fastest at every age except 3 months. For no age group was mean RT during the L-R sequence slowest, and at five of the nine assessments the average mean RTs were slowest during the IR sequence.

Mean RT as a function of age is plotted for each infant and for the group as a whole in Figure 10. The top left-hand panel of Figure 10a shows the group curve, which was obtained by averaging mean RTs across infants. Subsequent panels display individual infant data, that is, growth curves for mean RT across all trials for each infant at each age. Every infant showed a decrease in mean RT during the 2–12-month period (range = 102–231 ms). For most infants, age changes in RT are small from 2 through 4 months, followed by a period of rapid change between 4 and 8 months, after which the rate of change slows until 12 months.

When averaged across infants, mean RT drops an average of 155 ms,

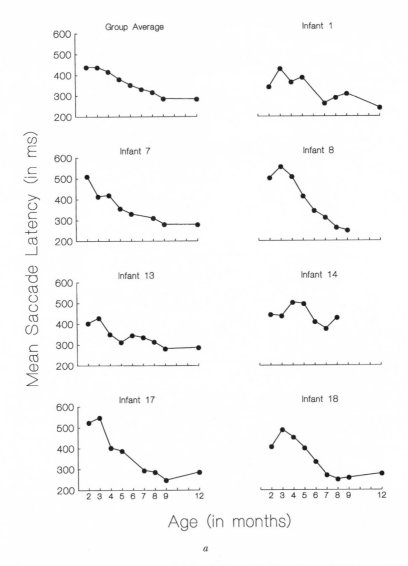

a

FIGURE 10.—Mean saccade latency (mean RT) as a function of age (2–12 months) for the group as a whole and for each infant.

from a mean of 440 ms at 2 months to 285 ms at 12 months. The decline in mean RT for the group appears to be monotonic; that is, for each succeeding month, mean RT is shorter than it had been the month before. However, these month-to-month differences tend to be small.

Our primary interest is in developing quantitative regression models for

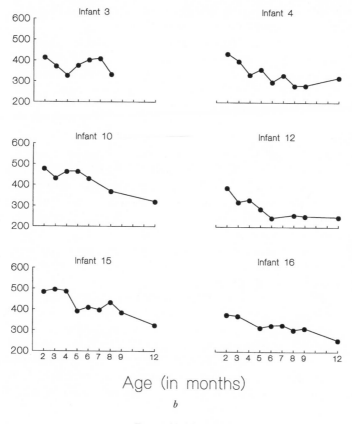

Age (in months)

b

FIGURE 10 (*Continued*)

growth, models that will describe growth at both the group and the individual levels. These models will be explored using both polynomial and truly nonlinear functions (i.e., models that are nonlinear in their parameters). Ideally, one could address nearly all questions about sequence type differences, group age trends, and individual differences in the context of a single random-regressions mixed model. Because there are three different sequences, this model should address the possibility that different functions or parameter estimates may be required to describe growth for each.

Therefore, the key variables included in the analysis would be sequence (three fixed levels), fixed regressions for the intercept, for linear change, and for quadratic change for month, and individual regressions expressed as deviations from the overall fixed regression on age (again, for the intercept and for the linear and quadratic terms). The intercepts and slopes of these individual regressions are random parameters in the model—as they must be from a sampling standpoint because they are associated with the levels of the factors for individuals, which are themselves random (Henderson, 1982). We also include a random effect for individuals themselves to capture individual differences not included in the random regressions for month. Furthermore, both the fixed and the random regressions can be looked at separately by sequence. Finally, the model should be estimated using an unstructured covariance matrix to provide for the possibility of covariances between individuals and between the intercept and the linear and quadratic parameter estimates at the individual level.

Unfortunately, we were unable to achieve convergence with restricted maximum likelihood (REML) or maximum likelihood (ML) estimation when we attempted to estimate the full model described above. In fact, we were unable to achieve convergence for any model that included parameters for either the linear or the quadratic terms at the individual level when sequence was included in the model. Furthermore, convergence did not depend on the choice of error structure (we tried simple and compound symmetry in addition to unstructured covariance matrix options). Therefore, we first present a model that includes fixed terms for sequence type and the linear and quadratic regression parameters separately by sequence type and that includes individual-specific intercepts. When we address individual differences in the shapes of growth curves in a later section, we show that, by simplifying the fixed-effects part of the model, it is possible to include regressions at the individual level, but only in linear form. However, our primary concern initially is with developing a growth model for the group that will also allow us to determine whether the different sequence types are associated with different patterns of growth.

Analyses of the mixed models were carried out using the SAS PROC MIXED procedure (SAS Institute, 1993). This prevented us from exploring truly nonlinear growth functions in the mixed-model context. Therefore, when we used the exponential growth model reported later in this chapter, we estimated individual functions directly using SYSTAT NONLIN (SYSTAT, 1992). This estimation was carried out separately for each individual in fixed-effects models.

The analysis revealed significant differences in mean RT as a function of sequence type ($F[2, 290] = 3.13$, $p < .02$). Pairwise comparisons revealed that only the IR and L-R sequences differed significantly ($t[290] = 2.43$, $p < .05$, Bonferroni adjusted). There were strong linear and quadratic trends across age ($F[3, 286] = 14.65$, $p < .0001$, and $F[3, 286] = 3.53$, $p < .002$, respectively), and these age trends were seen at each level of sequence type. The linear slope was large and significant separately for sequence IR (estimated slope = -35 ms/month, $t[286] = -3.90$, $p < .0001$), sequence L-R (estimated slope = -36 ms/month, $t[286] = -4.09$, $p < .0001$), and sequence L-L-R (estimated slope = -33 ms/month, $t[286] = -3.52$, $p <$

.0005). The linear rate of change did not differ among the three sequences ($F[2, 286] = .02$).

Although not as strong, the quadratic component of change was either significant or nearly so for each sequence type individually. Sequence IR had an estimated slope of 1.18 ms/month^2 ($t[286] = 1.83, p < .07$), and sequence L-R had the most rapid change, with an estimated slope of 1.39 ms/month^2 ($t[286] = 2.18, p < .03$). Finally, sequence L-L-R showed the smallest quadratic slope, with an estimate of 1.14 ms/month^2 ($t[286] = 1.62, p < .11$). Although there were slight differences in the magnitudes of the quadratic parameters, the contrast comparing these slopes across sequence types did not approach significance ($F[2, 286] = .04$).

We concluded from these analyses that there was considerable consistency in the pattern of age changes for the three sequences. RTs were slightly faster during the L-R alternating sequence, but the absolute magnitude of these differences was small (20 ms between IR and L-R, 17 ms between L-R and L-L-R; see Table 3).

With respect to our decision about what data to enter into subsequent analyses of age-related change and individual differences in mean RT, these findings do not suggest to us a clear rationale for carrying out separate growth-curve analyses, one for each sequence type. For each sequence we found a strong negative linear trend as a function of age but no differences between the sequences in this pattern. And we found significant quadratic trends but no differences between the sequences in this pattern either.

If, as suggested by these analyses, mean RT measures a similar attribute in all three sequences, then the intercorrelations among the mean RTs should be substantial. Because we were less interested in each of the 27 intersequence r's than in the overall pattern of intersequence correlation across all months, we computed the residual correlations in a mixed model that included age as a covariate. This analysis revealed substantial redundancy between the measures, as evidenced by a correlation of .35 ($p < .0012$) between mean RT in IR and L-R, .62 ($p < .0001$) between L-R and L-L-R, and .48 ($p < .0001$) between IR and L-L-R. We decided, therefore, that the increment to our understanding that might result from including the small amount of systematic variation associated with sequence type would be outweighed by introducing potentially confusing redundancies arising from three nearly identical analyses conducted on very similar data. Furthermore, by basing our analyses of RT on all the available data for each infant, we could obtain more stable estimates of infants' performance than if their scores were based on only a subset of the data. Consequently, in the RT analyses that follow, we ignore the distinction between the different sequences and treat the data as responses to a single series of stimuli.

At this point, it is important to clarify our assumptions regarding between-individual variation in growth curves relating RT to age. As mentioned

61

earlier, averaging individual growth curves can lead to serious distortions when characterizing the shape of the group developmental function, and averaging can prevent one from discovering distinctly different patterns of growth. This problem has been clearly documented in studies of physical growth. When there are individual differences in the growth rate for the timing of a growth spurt, the averaged growth curve does not accurately represent the shape of any individual growth function (Tanner, 1962, 1963; Wohlwill, 1973). It is also known that the magnitude of the problem of averaging across individuals depends to a considerable extent on the family of functions that underlies the growth process. Most relevant for our purposes is that, unlike polynomial functions, group exponential curves are particularly vulnerable to distortion when they are composed of an average of individual functions (e.g., Deming, 1957; Merrill, 1931).

With respect to RT, we were unwilling to assume that all individuals begin development at the same point or grow at the same rate, so we chose to model the growth functions of individual infants. To a large extent, the choice of fitting group or individual growth curves depends on whether one believes that individual differences represent only error variance or that they represent distinctly different types of individuals (Burchinal & Appelbaum, 1991). Previous research suggests that measured individual differences in saccade RT are not simply error variance. First, Canfield et al. (1995) reported a high degree of stability of individual differences for RT ($r = .81$) between the ages of 4 and 6 months. In addition, Haith and McCarty (1990) found stability in RT in their sample of 3-month-old infants. Moreover, findings from two prospective studies reported significant associations between infant saccade RT and later IQ (Benson et al., 1993; DiLalla et al., 1990), suggesting that individual differences in RT reflect some enduring attribute of individual infants. These facts suggest that the assumption that individual differences represent only error variance is not justified.

Since we are not assuming that "patterns of growth exhibited by all individuals within a population are identical" (Burchinal & Appelbaum, 1991, p. 26), we are not invoking "the strong concept of growth" (p. 25). But neither do we assume only a "weak concept of development" (p. 25) that allows each individual's growth data to be modeled by a unique function. We take an intermediate position, assuming that growth in RT for all individuals can be well described by a single family of functions (e.g., polynomial, exponential) but that the parameters may vary from one infant or group of infants to another.

Limiting ourselves to only one or a few functions to describe development for all individuals makes our models vulnerable to the obvious possibility that a family of functions other than the ones we consider could provide a better description of the growth of a particular infant, a subgroup of infants, or even the entire population (Stigler, Nusbaum, & Chalip, 1988). However, we were concerned that, without some self-imposed constraints, our efforts might become a "fishing expedition" for the best-fitting curve. Thus, in the description of models that follows, we constrained our search to parameterizations of a simple quadratic polynomial and an asymptotic exponential (Mitcherlitz) equation.

The polynomial function was chosen because most researchers are familiar with its properties, it is computationally more tractable than are exponential functions, and it requires few assumptions about the true shape of the underlying function being estimated (Burchinal & Appelbaum, 1991). The exponential function was chosen because it has been used to describe growth in many biological domains and because it has been applied to age-related changes in speed of information processing in particular (Kail, 1988, 1991c, 1993a). Although other models may fit the data as well or even better, we have no a priori reason to explore other models at this early descriptive stage in the research program.

In general, a random-regressions model tests for the significance of variance components under the assumption that individual variation is randomly distributed about the group mean. Although in principle such a model could include random regressions for the parameters of an exponential or some other nonlinear function, PROC MIXED is limited to testing for significant variance components in polynomial models. In addition, our random-regressions models testing for differences among sequences in the patterns of change over age would not converge when random factors for linear or quadratic change were included in this specific model. Partly in response to these constraints, we decided to carry out our individual growth-curve analyses by fitting curves directly to individual infant data. Using the NONLIN module in SYSTAT, we successfully estimated the parameters for both polynomial and exponential functions fit to the data from individual infants. Although they are not within a mixed-model framework, these models converge, and they allow us to compare the fits and plausibility of polynomial and exponential functions estimated using the same method.

In the following sections, we describe four models of individual growth for RT during the 2–12-month period. The first model is an estimation of growth-curve parameters using a quadratic polynomial function—first within an overall mixed model using PROC MIXED, then when fit separately for each individual using SYSTAT NONLIN. The remaining three models are different estimations using an asymptotic exponential function, all estimated using NONLIN. Each model will be evaluated to determine its plausibility for representing individual growth.

For our final model, after fitting the individual curves, we group together the curves with similar parameter estimates. From these groups, we construct three prototypical functions that we believe provide an accurate characterization of both the similarities and the individual differences in the nature of growth in this developmental domain.

Model 1: Individual Growth Curves Estimated by a Quadratic Polynomial

We estimated individual polynomial functions in two ways. First we reduced the complexity of our mixed model so that regressions were no longer estimated for sequence type (only age and age squared were estimated in the fixed part of the model). This change allowed the model to converge while including the intercept and age (linear) as random parameters. This was the only model that allowed us to estimate simultaneously the intercepts and slopes of the individual growth functions. Second, we estimated individual growth functions separately for each individual in NONLIN. The parameter estimates from these two models are then compared.

The mixed model that excluded sequence type revealed the same pattern of change at the group level (averaged over individuals) as found previously (when averaged over sequences). Not surprisingly, there was a significant linear trend of approximately the same magnitude as found for each of the individual sequences (estimated slope $= -35$ ms/month, $t[12] = -6.32$, $p < .0001$). Furthermore, the quadratic slope was very close to that found for the individual sequences in the previous model (estimated slope $= 1.27$, $t[280] = 3.37$, $p < .001$).

In addition to estimating parameters at the group level, this model estimates parameters for intercept and linear slope for each infant under the random part of the model. We can use these estimates to test for significant variability among individuals, under the assumption that the individual parameters are normally distributed about the estimate from the fixed-effects part of the model. The results of this analysis were somewhat equivocal. A significant variance component was found for the intercept ($Z = 2.10$, $p < .04$), and the variance estimate for the linear slope was marginally significant ($Z = 1.73$, $p < .09$). Given the small number of infants included in this analysis, it is surprising to find significance in these variance components, suggesting the presence of different developmental starting points and possibly different trajectories. Recall that, in this model, we were unable to estimate quadratic slopes for individuals and that this will have influenced our estimation of the other two parameters. Thus, in order to describe simultaneously the pattern of individual growth and the variation among individuals in this pattern, we require a more detailed approach.

Describing Polynomial Growth in Individuals

We wanted to include as much data as possible in our mathematical descriptions of individual growth, and we wanted our models to be based on the most informative data we had available. Thus, we chose not to use the global mean RT as the dependent measure for the following models. The mean RT provides only a single data point for each assessment, and it carries with it the vulnerability that mean RT for an assessment based on only a few observations will have the same weight in the analysis as does mean RT for an assessment based on many. This would be undesirable because the number of usable trials differed substantially from one month to the next for individual infants. Constructing a model based on the three sequence means would provide a greater number of observations, but those means themselves are based on very different numbers of individual trials within an infant. An obvious method for including as much data as possible, without introducing bias from differing numbers of observations both within and between sessions, is to use the individual trial data. This method carries the benefit of automatically weighting the data in proportion to the number of trials for any given assessment.

Our preliminary analyses revealed that the raw data were too variable to allow for precise modeling of age trends. Consequently, under the assumption that much of the intertrial variability within sessions was due to measurement error, we used a smoothing algorithm to reduce the noise (Diggle, 1990). Because our goal is to discover the nature of the growth function, we needed to avoid using any algorithm that carries with it assumptions about the nature of the underlying functional form. Using linear or exponential smoothers, for example, might lead us to misattribute the resulting functional form of the growth curve to the nature of development when it might in fact be due to an interaction between the true shape of the growth function and the smoothing algorithm. Therefore, we used the LOWESS smoothing algorithm available in SYSTAT (tension = .75; Chambers, Cleveland, Kleiner, & Tukey, 1983). LOWESS is a locally weighted least-squares smoother that makes no assumptions about the form of any underlying functions.

At this point, we also dealt with two anomalies in the data that needed to be addressed to meet the assumption of monotonicity (required for our subsequent nonlinear analysis). First, the growth curves from two infants were clearly not monotonic. Both infant 3 and infant 14 evidenced profound non-monotonic change that could not be explained by uncharacteristic data for only a single assessment (see Figure 10). Data from these two infants are not included in the subsequent analyses of growth in RT. Second, we needed to address a local nonmonotonicity that was consistent but quite unexpected. An inspection of Figure 10 shows that a subset of infants had a faster mean RT at 2 months than at 3 months. This finding conflicts with previous studies comparing average RTs from groups of 2- and 3-month-old infants, for which a large drop in mean RT between these ages has been found (e.g., Canfield & Haith, 1991; Wentworth & Haith, 1992).

In the present data set, it seems that this anomaly results, not from unusually long latencies at 3 months, but rather because our infants' average latencies at 2 months are unusually short. Whereas the mean RT from studies of 2-month-olds reported in Table 1 above is nearly 600 ms, in our sample it was only 440 ms. Thus, infants 1, 8, 13, 17, and 18 were approximately 100–200 ms faster at 2 months than has been previously reported. In comparison, their mean RTs at 3 months were much closer to the average of 500 ms reported in Table 1 for previous studies (see Figure 10).

Our speculations about the explanation for this curious finding suggest that it may result from an interaction among age, the development of postural control, and stimulus properties. We noted earlier that our stimuli are larger and brighter than the stimuli used in previous studies. The images are also animated, and our infants were sitting up rather than lying on their backs. Given the profound developments in eye-head coordination that occur during the 2–4-month age period (Goodkin, 1980; Regal, Ashmead, & Salapatek, 1983), it is possible that these stimulus factors interacted with the development of postural control and seating position.

For the least-mature 2-month-olds, maintaining a seated position likely required a great deal of physical support from their parent. Some of these infants may have rested their heads back on their parent's chest and relied exclusively on eye movements to shift their gaze back and forth. We postulate that, for these infants, the highly intense stimuli, and possibly the upright position, may have led to the uncommonly fast latencies we observed. Alternatively, the more mature 2-month-olds, like the 3-month-olds who had slightly better postural control, may have been struggling to control their posture by themselves—attempting to coordinate their head, neck, and eye movements to shift fixation. The increased effort required to coordinate the head and neck may have made them slower to shift fixation with their eyes. Such an apparent regression in development would not be exceptional. Thus, some of our data from 2-month-olds may represent a different viewing strategy, suggesting the possibility of different age trends depending on stimulus intensity and whether head control is required.

Unfortunately, our recording system does not provide us with the information we need to test this conjecture. But, given the consistency of previous research that reports a substantial decline in RT from 2 to 3 months, we found it difficult to believe that, when measured under comparable conditions, the general trend in RT between 2 and 3 months is in the direction of older infants being slower. Therefore, we decided to use data beginning at each infant's maximum RT, whether it occurred at 2 months or at 3 months of age. Our results should be interpreted with the possibility in mind that slightly different procedures may result in different estimated growth functions.

After smoothing each infant's individual trial data, we returned to our modeling in an effort to reveal the underlying shape of the individual developmental functions, beginning with a three-parameter polynomial function.

TABLE 4

MODEL 1: PARAMETER ESTIMATES AND R^2 FOR INDIVIDUAL POLYNOMIAL
GROWTH FUNCTIONS FOR MEAN RT

Subject	$\hat{\phi}_1$	$\hat{\phi}_2$	$\hat{\phi}_3$	R^2
No. 1	517	−48	2.19	.965
No. 4	455	−30	1.20	.969
No. 7	536	−49	2.21	.965
No. 8	927	−132	6.22	.971
No. 10	485	−12	−.09	.927
No. 12	427	−37	1.83	.967
No. 13	412	−20	.67	.942
No. 15	563	−33	1.30	.982
No. 16	395	−16	.31	.961
No. 17	700	−84	3.99	.961
No. 18	806	−108	5.2	.972
Median	517	−37	1.83	.965

NOTE.—In all cases, we report the corrected R^2 (1 − residual/corrected). ϕ_1 represents the y intercept, ϕ_2 the linear slope, and ϕ_3 the quadratic slope.

Equation (1) shows such a model, where ϕ_1 represents the y intercept, ϕ_2 the linear slope, and ϕ_3 the quadratic slope:

$$Y = \phi_1 + \phi_2 X + \phi_3 X^2. \tag{1}$$

Individual quadratic growth functions based on Equation (1) were fit to each infant's RT data using a least-squares nonlinear modeling program (SYSTAT, 1992). The simplex method was used to minimize the loss function. Models were successfully estimated for all 11 infants remaining in the analysis, and all models converged within 16 iterations. Table 4 shows the estimated intercept and slope parameters for each child as well as the estimated index of fit (R^2).

These models can be further evaluated by inspecting the model fits to the raw data and the pattern of residuals. In Figure 11, each panel contains three plots for each infant, plotted on a single x-y axis. The upper curve of each panel shows the smoothed individual trial data (solid bars) overlaid with the growth curve estimated from the data by Equation (1) (solid line). The lower plot of each pair is a residual plot that shows how the model residuals are distributed around zero. Figure 11 shows that the estimated polynomial function generally passes through the center of the smoothed data, and in nearly every case the residual values are normally distributed around zero. R^2 ranged from 92.7% to 98.2% for individual infants, indicating that the polynomial function provides an excellent fit to the smoothed data. On average, Equation (1) accounted for 96.5% of the variance in RT (see Table 4).

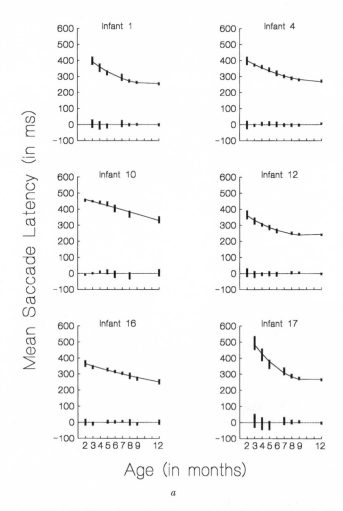

FIGURE 11.—Model 1: Plots of smoothed data, model prediction, and residuals for three-parameter quadratic polynomial. All parameters are free. Each panel contains three plots for each infant. The upper curve of each panel shows the smoothed individual trial data at each age (solid bars) overlaid with the growth curve (single line) that was estimated from the data by Equation (1). The lower plot of each pair shows model residuals. Although the residuals and smoothed data appear as solid bars, in fact they are densely overplotted symbols. Each symbol represents a single reactive saccade latency. Ends of bars correspond to the range of the *smoothed* data series for a given assessment. Numbers of RTs vary for each infant and month.

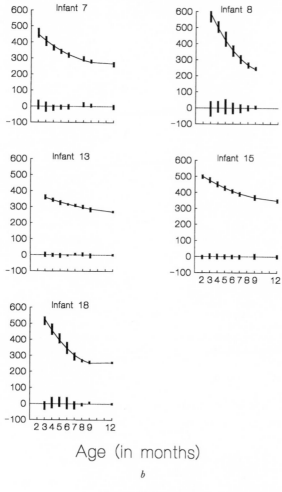

Age (in months)

b

FIGURE 11 (*Continued*)

Model 1 Evaluation

The polynomial function provides a simple and readily understood model for describing the shape of the growth curve in RT during the 2–12-month period. The assumptions required by the model are easily met, and, in the absence of a theory to characterize the nature of the growth process, we would be likely to end our quantitative description of individual growth at this point. Nevertheless, despite all its strengths and simplicity, the model suffers from several important limitations. By constructing a group function from the median parameter estimates of the individual functions and then

69

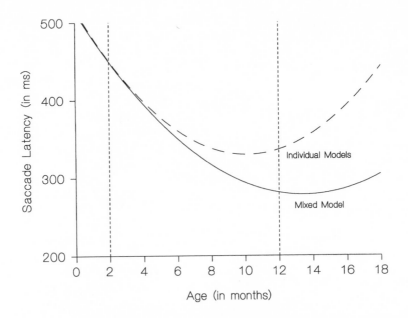

FIGURE 12.—Average individual growth curve predicted by Model 1 (polynomial) for ages 0–18 months. For the dashed curve, parameters are median parameter estimates from the individual growth models (Equation [2]). For the solid curve, parameter estimates are derived from the mixed-model analysis. Vertical dashed lines represent the range of the actual data values (2–12 months) from which the parameters are estimated.

plotting it for values both inside and outside the range of our data, we can easily see how this particular model fails. This function represents age changes in RT as

$$RT = 517 - 37 \text{ (age)} + 1.83 \text{ (age}^2). \tag{2}$$

That is, the average function has an estimated y intercept of 517 ms and declines linearly by 37 ms/month, with a 1.83-ms quadratic slope. Figure 12 shows the predicted average growth curve as a plot of saccade latency by age beginning at birth (age = 0) and ending at 18 months. This figure shows both the quadratic model estimated from the individual trial data (solid line) and the curve estimated under the mixed-model approach (dashed line). The mixed-model curve was constructed from the fixed-effects parameter estimates. Note that the model derived from individual parameter estimates seriously underestimates the magnitude of the decline in RT during the second half of the first year. The curve based on the mixed-model parameter estimates provides a more plausible fit to the data, but both models make the improbable prediction that at some point infants begin to get slower, with the rate of increase in RT equal to the rate of decline predicted during the first 12 months. Stated more simply, the model suffers from the use of a quadratic term that produces a function in which the rate of decline is inevitably matched by the rate of subsequent increase.

Known facts about development contrast with the predictions from the quadratic models. We know that RT continues to decline after the age of 10 months, eventually reaching an adult mean of about 150 ms under the type of viewing conditions used in our study (Findlay, 1981; Shea, 1992). In addition, when Canfield, Wilken, and Schmerl (1991) studied saccade RT in a group of 4–6-year-old children (using the same stimulus sequence shown to infants in the present study), they found that RT averaged 211 ms.

Although the polynomial growth function is convenient to estimate, it also suffers from several more general limitations. First, polynomial functions provide implausible mathematical representations of a broad range of biological growth processes. Although U-shaped functions have been used sporadically to model growth, we suspect that growth in RT is produced by cumulative growth processes—which are nearly always better described by inherently nonlinear functions (Bogin, 1988). In addition, although the quality of the fit to the data provided by polynomial functions may be as good or better than the fit of more plausible inherently nonlinear models involving exponential or logistic functions (Burchinal & Appelbaum, 1991), these latter functions are more likely to provide meaningful parameter values (e.g., parameters representing initial performance level, asymptotic performance level, and the rate at which asymptotic performance is attained). Finally, as we discovered, a general characteristic of polynomial growth-curve approximations is that, even when they provide an acceptable statistical fit in the age range of the modeled data, polynomial models often become implausible when used to predict growth beyond the range of the data. In the present study, even the model based on the mixed-model parameter estimates became implausible when used to extrapolate beyond about 14 months.

Inherently Nonlinear Growth Functions

Inherently nonlinear functions are typically used to model processes of biological growth (Bogin, 1988; Burchinal & Appelbaum, 1991; Wohlwill, 1973), and the exponential function in particular has been applied to the development of processing speed (Kail, 1991c). Developmental changes in processing speed have been explained by two major approaches. One approach emphasizes cognitive processes in specific skill domains (Chi, 1977a, 1977b; Roth, 1983), while the other focuses on global changes (Kail, 1991b, 1993a). The specific-skills approach attributes age-related changes in processing speed to changes in specific processes, domains, or tasks. Examples of these changes include the development of more efficient strategies, the

71

elaboration of knowledge in specific domains, and a shift from the effortful use of algorithms to the more automatic direct retrieval of appropriate responses (Canfield & Ceci, 1992; Chi & Rees, 1983; Stigler et al., 1988). In contrast, the global hypothesis proposes that a significant portion of the developmental change in processing speed reflects an underlying mechanism that is not specific to any particular task (Kail, 1991b, 1993b).

Kail (1988, 1991c, 1993a) gathered evidence relevant to the global trend hypothesis by analyzing group developmental functions associated with a variety of perceptuomotor and cognitive tasks. Using a cross-sectional design, Kail obtained multiple measures of processing speed in children and adults ages 7.5–43 years. The tasks ranged in complexity from simple reaction time, to name retrieval, to measures of the time required to judge the accuracy of simple arithmetic problems. Kail found that RT declined throughout the age range for each task. Furthermore, he determined that the shape of the growth function was nearly identical for every task. Performance on four of the six speed tasks was well described by the exponential equation

$$Y = \theta_1 + \theta_2[e^{(-\theta_3 \times \text{age})}]. \tag{3}$$

In Equation (3), θ_1 represents the processing time required by a mature adult, $\theta_1 + \theta_2$ defines the y intercept, and θ_3 represents the rate of change in speed with respect to age (the exponential decay parameter). Kail (1991c) reasoned that the global trend hypothesis would be supported if age changes on all tasks declined at the same rate, whereas the specific-skills hypothesis would be supported if different decay parameters were needed to represent age changes for different tasks. Thus, in one analysis, he fixed the parameter representing adult asymptotic performance at the mean of an adult control group, and he fixed the decay parameter at $\theta_3 = .334$ (a value based on his previous research using visual and memory search tasks; Kail, 1988). This left only θ_2 to be estimated from the data.[10] This parameterization of the exponential function accounted for nearly 90% of the variance associated with age-related change for four of the six tasks. These findings are consistent with the conclusion that processing time declines exponentially at a rate common to many distinct psychological processes. Kail (1991b) argues that the findings support the hypothesis that "an exponentially changing global mechanism limits speeded performance" (p. 166).

Because Kail relies on cross-sectional data, any straightforward interpretation of his growth models requires the assumption that development proceeds in nearly the same manner in all individuals—which implies an endorsement of the strong concept of growth. As Kail himself recognizes, his

---

[10] In more recent reports (e.g., Kail, 1993a), Kail has revised his estimate of the universal decay parameter to $\theta_3 = .21$. But note that the fit of this type of function is insensitive to relatively large differences in θ_3.

TABLE 5

MODEL 2: PARAMETER ESTIMATES AND R^2 FOR INDIVIDUAL EXPONENTIAL
GROWTH FUNCTIONS FOR MEAN RT

Subject	$\hat{\theta}_1$	$\hat{\theta}_2$	$\hat{\theta}_3$	R^2
No. 1	−148	583	.036	.857
No. 4	−284	713	.026	.936
No. 7	252	352	.275	.975
No. 8	80	919	.199	.969
No. 10	−263	755	.020	.924
No. 12	−111	482	.032	.828
No. 13	−130	522	.025	.927
No. 15	316	275	.163	.968
No. 16	−119	511	.028	.959
No. 17	−221	778	.047	.843
No. 18	219	809	.314	.952
Median	−119	583	.036	.936

NOTE.—In all cases, we report the corrected R^2 (1 − residual/corrected). θ_1 represents asymptotic performance, $\theta_1 + \theta_2$ the y intercept, and θ_3 the rate of exponential decay.

descriptions of developmental functions for processing speed "may not reflect patterns of individual growth" (1991c, p. 264), and he cites evidence of possible sex differences in the parameters of the growth function. Because it is not possible to characterize the nature of growth in the individual in the absence of longitudinal data, it is possible that Kail's model is appropriate for describing age changes in processing speed only at the group level, and it is unclear what would be the meaning of only aggregate regularity. Nevertheless, if we accept the assumptions of the strong concept of growth, we can consider his research findings to suggest a specific alternative to the polynomial growth function.

Model 2: Individual Growth Curves Estimated by an Asymptotic Exponential

Under the assumption that developmental change in processing speed is to some extent due to exponential change in a global mechanism, we hypothesized that age changes in saccade RT during infancy should be well described by Equation (3). In addition, an exponential model appears to be a reasonable choice given that visual inspection of the individual growth curves suggests evidence for an asymptote in the age-related decline in mean RT beginning at around 9 months.

Using the same methods of model estimation as for the quadratic function, we fit the three-parameter exponential function described by Equation (3) to the smoothed RT data separately for each infant. Note that, for this model, we estimated all three parameters from the data. The exponential function was successfully estimated for each individual. Table 5 contains the

73

parameter estimates and R^2 values for individual infants, with median values in the bottom row. The fit was acceptable for most infants (median R^2 = .936, range = .828−.975).

Figure 13 shows individual model-fitting results for Equation (3). Looking across the panels corresponding to each infant, one can see substantial variation among the individual growth functions. In the case of several infants (e.g., infants 1, 4, 12, 13, 15, and 18), the model residuals indicate a consistent underestimation of the RT values for the 12-month data. To a lesser extent, this model also underestimates RT at 2 months (e.g., infants 1, 12, 13, and 15).

Model 2 Evaluation

The parameter values for asymptotic performance estimated for Model 2 are implausible because they predict that most infants will attain negative latencies at asymptote (see estimates for θ_1 in Table 5). In addition, the average individual function derived from median parameter estimates indicates that infants will reach mature performance when they are only about 24 months old (model prediction of RT at 24 months = 155 ms). In addition, many of the estimated functions appear to show only linear change as a function of age, suggesting that this exponential model may not be appropriate for all infants.

Our difficulty accurately describing the growth function using Equation (3) may arise from several sources. For example, it might result from the nature of our data set, specifically, from the fact that our measurements were taken only during the very early part of the developmental period. Our data represent the most rapid and probably the most variable period of growth. This could seriously limit our ability to estimate asymptotic performance because we have no data at near-adult performance levels to constrain our function. It is also possible that no single function will fit the growth pattern for the entire developmental period. This would suggest the presence of a discontinuity in the growth function sometime between 12 months and maturity, possibly reflecting the addition or deletion of certain underlying growth processes.

We reasoned that, if our difficulty in achieving a plausible fit to the individual growth functions is due to the absence of data at the asymptote, then we should observe better fits and better-behaved residuals if we fix θ_1 at the known adult average RT acquired under similar experimental conditions. If the parameter estimates remain implausible or the fit is no better than we found with Model 1, then we will consider the possibility that more than one function is needed to describe growth through maturity.

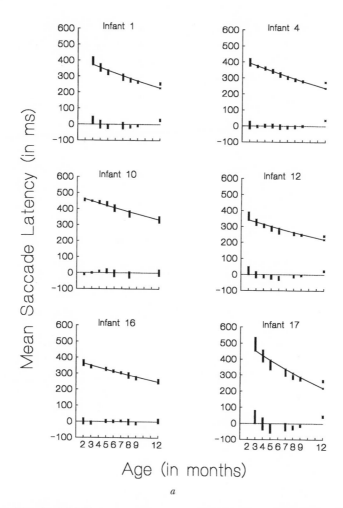

FIGURE 13.—Model 2: Plots of smoothed data, model prediction, and residuals for three-parameter exponential function. All parameters are free. Each panel contains three plots for each infant. The upper curve of each panel shows the smoothed individual trial saccade RT data (solid bars) overlaid with the growth curve (single line) that was estimated from the data by Equation (3). The lower plot of each pair shows model residuals. Although the residuals and smoothed data appear as solid bars, in fact they are densely overplotted symbols. Each symbol represents a single reactive saccade latency. Ends of bars correspond to the range of the *smoothed* data series for a given assessment. Numbers of RTs vary for each infant and month.

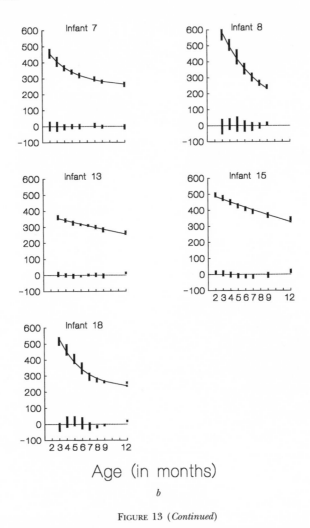

Age (in months)

b

FIGURE 13 (*Continued*)

Model 3: Fixing the Asymptote (θ_1) to the Known Adult RT Value

Research by Findlay (1981) includes experiments in which adults make saccades under viewing conditions similar to those used in the present study. Findlay's findings suggest an asymptotic saccade RT of about 150 ms. Therefore, in Model 3, we fixed θ_1 at this adult value and estimated only θ_2 and θ_3. Table 6 contains parameter estimates and associated R^2 values for this model. Plots of observed and predicted curves and their associated residual plots are shown in Figure 14.

TABLE 6

MODEL 3: PARAMETER ESTIMATES AND R^2 FOR INDIVIDUAL EXPONENTIAL
GROWTH FUNCTIONS FOR MEAN RT WITH THE ASYMPTOTE SET AT THE
ESTABLISHED ADULT LEVEL (150 ms)

Subject	$\hat{\theta}_2$	$\hat{\theta}_3$	R^2
No. 1	333	.116	.914
No. 4	297	.091	.958
No. 7	364	.116	.937
No. 8	947	.255	.967
No. 10	349	.052	.916
No. 12	247	.104	.891
No. 13	254	.068	.937
No. 15	390	.059	.951
No. 16	250	.074	.960
No. 17	520	.157	.917
No. 18	680	.203	.931
Median	349	.104	.937

NOTE.—In all cases, we report the corrected R^2 ($1 -$ residual/corrected). $150 + \theta_2$
represents the y intercept, θ_3 the rate of exponential decay.

As can be seen in Table 6, one effect of fixing the asymptote was to increase the magnitude of the decay parameter for seven of the eleven infants. For six of those seven, Model 3 provided a better fit to the data (as indicated by R^2) than Model 2. There were four infants whose decay parameters declined when θ_1 was set at 150 ms, and for all of those infants R^2 values are smaller for this model than for Model 2. There was one infant (infant 8) for whom the decay parameter increased but the R^2 did not, and, in her case, R^2 remained high, declining by only .002. Median R^2 for this model was .937 and ranged from .891 to .967. Thus, the net result of fixing the asymptote was that the median R^2 for Model 3 was essentially unchanged from Model 2 (although the mean R^2 increased from .922 to .934). Nevertheless, by fixing one parameter, this model has accomplished with only two parameters what the previous model explained with three. In addition, having the asymptote fixed at the known adult value increases the plausibility of the model because no infant is predicted to reach asymptote at a value below what adults actually achieve (or even below zero).

Model 3 Evaluation

The modest superiority of the second exponential model to the first does not necessarily make it satisfactory. An important shortcoming of the third model is that the average individual function (based on median parameter

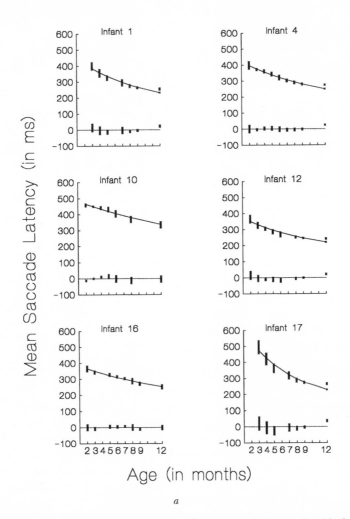

Age (in months)

a

Figure 14.—Model 3: Plots of smoothed data, model prediction, and residuals for the three-parameter exponential function. θ_1 is fixed at 150 ms; remaining parameters are free. Each panel contains three plots for each infant. The upper curve of each panel shows the smoothed individual saccade RT trial data (solid bars) overlaid with the growth curve (single line) that was estimated from the data by Equation (3). The lower plot of each pair shows model residuals. Although the residuals and smoothed data appear as solid bars, in fact they are densely overplotted symbols. Each symbol represents a single reactive saccade latency. Ends of bars correspond to the range of the *smoothed* data series for a given assessment. Numbers of RTs vary for each infant and month.

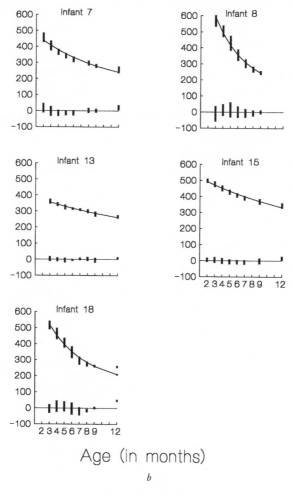

Age (in months)

b

FIGURE 14 (*Continued*)

estimates) reaches its asymptote at a far younger age than existing data would suggest is plausible. Specifically, the model predicts that, by the age of 2.5 years, infants will have reached nearly asymptotic levels of performance (predicted RT at 30 months = 158 ms).

We are aware of only one study reporting saccade latencies for children in this general age range. In a study of saccade RT, attention, and mental processing in young children (age range = 3.5–7 years, mean age = 5 years) using the same stimulus sequence shown to babies in the present study (and a minimum RT set at 133 ms), Canfield et al. (1991) found an average group

79

RT of 217 ms. This is substantially slower than what the model predicts for 5-year-olds using the current parameter estimates (predicted RT at 60 months = 150.2 ms).

Predictions from this model are also inconsistent with research using somewhat different RT paradigms that report a substantial decline in mean saccade latency during childhood (Groll & Ross, 1982). For example, Ross, Radant, Young, and Hommer (1994) reported a monotonic decline in saccade RT from 8 to 15 years of age, from a mean of nearly 220 ms to almost 185 ms. In our own study, in addition to conflicting with known facts about when asymptotic performance is attained, Model 3 also underestimates RT for our 2-month-olds by approximately 25 ms, and it underestimates our 12-month mean RT by approximately 30 ms (see the residual plots in Figure 14).

Given that the parameter estimates from the two previous models specified individual growth curves that either reached an implausibly fast asymptote or, more commonly, reached their asymptotes far too quickly, we explored a third parameterization of Equation (3). We hypothesized that one reason the previous models produce unsatisfactory results is that there may be no single function that adequately represents age-related change in saccade RT across the entire developmental period. We observed that many infants seemed to reach a local asymptote, leading us to suspect that our data may reflect development during a unique early period of growth and that one or more additional functions would be needed to describe age-related change for subsequent growth periods.

Also instructive were the results that we obtained when we fixed the asymptote at 150 ms. In relation to model fit, there was a differential effect of fixing this parameter for different groups of infants. For all but one of the eight babies whose estimated asymptote from Model 2 (estimating all three parameters) was below 150 ms, their R^2 was substantially raised by fixing the parameter. For the three infants whose estimated asymptote from Model 2 was above 150 ms, R^2 was substantially lowered when the asymptote was fixed. Together with the fact that the median R^2 was unchanged, these findings suggest to us that the asymptote may be higher than 150 ms—which is consistent with the hypothesis of a local asymptote at a substantially slower RT than that achieved by adults.

Although our data do not tell us precisely at what performance level or at what age the hypothesized local asymptote occurs, our sample of infants includes several who reached a somewhat steady performance level as early as 7 or 8 months—and maintained it through 12 months. Interestingly, the slowing of age-related change seemed to be linked to performance (a mean RT shorter than 300 ms) rather than to age (age ranged from 5 months to 12 months or older). Consequently, we defined asymptotic functioning solely in terms of an RT criterion.

TABLE 7

MODEL 4: PARAMETER ESTIMATES AND R^2 FOR INDIVIDUAL EXPONENTIAL
GROWTH FUNCTIONS FOR MEAN RT WITH THE ASYMPTOTE SET AT THE
HYPOTHESIZED LOCAL ASYMPTOTE OF 242 MS

Subject	$\hat{\theta}_2$	$\hat{\theta}_3$	R^2
No. 1	392	.309	.968
No. 4	230	.186	.962
No. 7	346	.246	.974
No. 8	1,159	.394	.943
No. 10	264	.081	.903
No. 12	276	.398	.972
No. 13	188	.156	.943
No. 15	311	.099	.973
No. 16	176	.171	.937
No. 17	724	.352	.969
No. 18	904	.372	.947
Median	311	.246	.962

NOTE.—In all cases, we report the corrected R^2 (1 − residual/corrected). 242 + θ_2 represents the y intercept, θ_3 the rate of exponential decay.

Model 4: Fixing (θ_1) at the Hypothesized Local Asymptote for RT

We made the assumption that our fastest infant at 12 months had reached asymptote. Further, we assumed that the absolute value of the asymptote differs little among infants. We then estimated individual growth functions for which we had fixed the asymptote at the fastest infant's 12-month mean RT (θ_1 = 242 ms) for all individuals. We hypothesized that, if development in saccade RT from early infancy is exponential in form, but only down to a local asymptote, then we should observe generally better model fits and more normally distributed residuals when the asymptote is fixed at that value. However, if generally less adequate models are the result, then we will need to reexamine one or more of our assumptions.

Model information, presented in Table 7, reveals noticeably higher R^2 values (median = .962, range = .903–.974) and substantially larger decay parameters (median = .246, range = .081–.398) than were found for the previous two model equations. In addition, the y intercept increased from a median of 499 ms for the previous model to 553 ms for the present model. Furthermore, this model provided the best (or very nearly the best) fit for eight of the 11 infants. Figure 15 reveals this general improvement in fit. Nearly every estimated growth function passes through the center of the smoothed data, and the residual plots are nearly all symmetrical about zero. With the exception of infant 8, who shows a steep decline through 8 months but did not return for the 12-month assessment, there is little under- or overestimation of RT at any age.

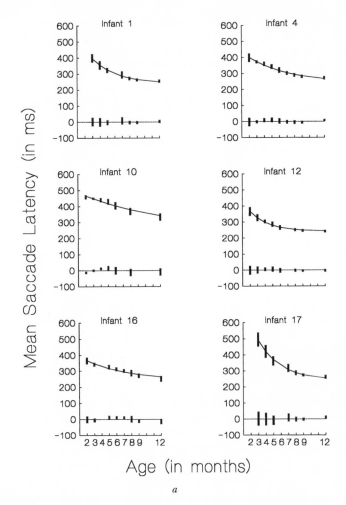

Age (in months)

a

FIGURE 15.—Model 4: Plots of smoothed data, model prediction, and residuals for the three-parameter exponential function. θ_1 is fixed at 242 ms; remaining parameters are free. Each panel contains three plots for each infant. The upper curve of each panel shows the smoothed individual trial saccade RT data (solid bars) overlaid with the growth curve (single line) that was estimated from the data by Equation (3). The lower plot of each pair shows model residuals. Although the residuals and smoothed data appear as solid bars, in fact they are densely overplotted symbols. Each symbol represents a single reactive saccade latency. Ends of bars correspond to the range of the *smoothed* data series for a given assessment. Numbers of RTs vary for each infant and month.

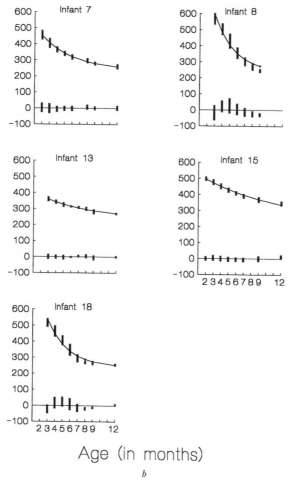

Age (in months)

b

FIGURE 15 (*Continued*)

Model 4 Evaluation

We believe that the two-parameter exponential with the local asymptote set at 242 ms provides the most plausible parameter estimates and the best fit to the data of the three exponential models considered thus far. Using median parameter estimates to derive an average individual growth curve, the model predicts that (local) asymptotic RT will typically be reached by the age of 16–18 months. The improvement in model fit when the asymptote is fixed to a value that is achieved during infancy suggests that either an additional function or a different parameterization of the present function will be needed to describe growth in saccade RT beyond the first year of life.

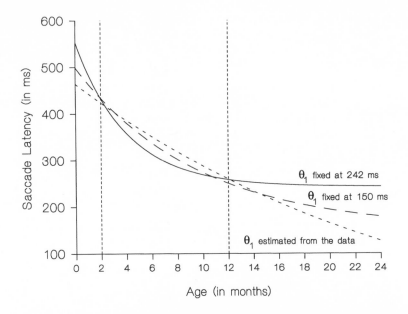

Figure 16.—Average individual growth curves for infant saccade latency as predicted by Models 2–4 for ages 0–24 months. Parameters are median parameter estimates from the individual exponential growth models (Equation [3]). Vertical dashed lines represent the range of the actual data values (2–12 months) from which the parameters are estimated.

There exists one study of saccade RT in older children that supports our conclusion that a different growth model may be needed for the period of childhood. In a cross-sectional study, Ross et al. (1994) found that group average saccade RT declined in a linear manner during the 8–15-year period, and there was no indication of an exponential trend in the data. However, because they did not study individual growth, the data can be considered only suggestive of what might be found in a longitudinal study.

Summary and Evaluation of the Three Exponential Models

Average individual developmental functions constructed from the median parameter estimates for each of the exponential models are shown in Figure 16. The implausibility of Model 2 (θ_1 estimated from the data) is apparent in that it predicts that infants will reach the adult asymptote by 24 months of age. Model 3 was constrained to reach the known adult asymptotic value of 150 ms, but with a single exponential function. Overall, this model fits as well as Model 2, while using one parameter fewer. However, it also makes an implausible prediction outside the range of the data—that adult-level performance will be reached by the age of 30–36 months. Under the assumptions that (1) our data are representative of the true growth function, (2) there is a single function that describes growth from infancy to maturity, and (3) this function has the form of the asymptotic exponential (Equation [3]), Model 3 (θ_1 fixed at 150 ms) should have provided the best fit overall, and it should have made accurate predictions when extrapolating beyond 12 months. However, the superiority of Model 4 suggests that one or more of these assumptions are false.

We believe that the data from at least some of our infants represented a good estimate of the true shape of the developmental function and that this function was well described by Equation (3). However, the best fit to the data was found when the model was constrained to reach only a local asymptote. These findings suggest that the assumption of a single growth curve from infancy into adulthood must be questioned. If more than one growth function is needed to describe development in RT during infancy and childhood, then future research should work to identify what growth processes are added or deleted that may account for the change in the pattern of growth.

INDIVIDUAL DIFFERENCES IN RT

Individual Differences in the Shape of the Growth Function

As several writers have commented, the study of individual differences in development is restricted almost exclusively to the reporting of cross-age correlations that represent the degree to which individuals maintain their rank order on a measure over a set of repeated assessments (Appelbaum & McCall, 1983; McCall et al., 1977; Wohlwill, 1973). But individual differences may also exist in the form of the growth function. That is, individuals may differ in terms of an intercept value, the rate of change with age, or the final level of performance they reach. By considering individual differences in the nature of the growth function, a frequently condemned division in developmental psychology between studies of normative growth and studies of individual differences is undone. And, as our description of the "normative" shape of the growth curves has already revealed, individual differences are an essential part of that understanding. These differences can be elaborated on further.

First, recall that we have already used a random-regressions model to test for significant variation in the intercept and linear slope among infants. Although this model is suggestive of the types of differences we may find, it was seriously compromised because it was not possible to test for individual differences in the parameters of the exponential function, which we have argued provides a better description of individual growth. Thus, we address

85

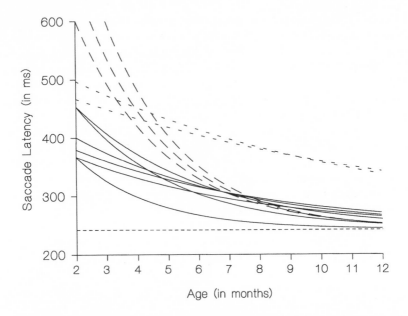

FIGURE 17.—Individual exponential growth functions for infant saccade latency using parameter estimates from Model 4. Dashed horizontal reference line is at the fixed asymptote (θ_1 = 242 ms).

the issue of individual differences in the shape of the exponential using a more descriptive approach.

Although we were satisfied with the degree to which Equation (3) was able to fit the individual growth data for all infants, there were still noteworthy differences in the shapes of the growth curves and parameter estimates among the infants. As shown in Figure 17, the individual growth curves for Model 4 (θ_1 fixed at 242 ms) fall into three somewhat distinct groups. In the largest group (plotted with solid lines) are six infants showing relatively low y-intercept values and a modest exponential bend. In the next largest group (plotted with the longer dashed lines) are three infants with very high y intercepts and very rapid exponential decline. Although these three infants are much slower in the first few months than infants in the other two groups, their sustained rapid growth rate leads them to be three of the fastest infants at 12 months. In the smallest group (plotted with the shorter dashed lines) are two infants with moderately high y intercepts and almost no exponential decline. These infants never reached a mean RT shorter than 325 ms at any age and seem to be on a distinctly different developmental trajectory.

It is important to emphasize that, although the number of infants is small, a great deal of information has gone into the estimation of these growth functions. Each individual growth function is based on a consistent pattern of responding over many months and typically estimated from more than 500 data points. It would not be possible for an infant's current group membership to be changed on the basis of having more fast or slow RTs for only one or two assessments; rather, the differences between groups represent differences in responding that persisted across much of the first year of life. Future studies of larger samples of infants may be able to test quantitatively for the presence of multiple growth functions using some of the techniques described in Burchinal and Appelbaum (1991).

Stability of Individual Differences across Age

The nature of the individual differences noted above resides in differences in the patterns of change across age. However, considerable practical and theoretical value is attached to documenting and explaining the stability and instability of individual differences. When individuals in a group tend to maintain their relative rank on a measure across a considerable age range, it is often an indication of an underlying continuity in the person-environment interaction being measured. It is important to mention here how the description of individual differences may supplement the analysis of stability. Referring back to Figure 17, if one believes that the differences between the growth curves for this small sample of infants are representative of the population of such growth curves generally, then one might find it useful to study stability within particular groups of infants characterized by the shapes of their growth curves.

Because findings of substantial predictive correlations between infant cognitive performance and childhood intelligence test performance have been widely replicated (Bornstein & Sigman, 1986; Colombo, 1993; Rose et al., 1988), the nature of individual differences in infant information processing has become an exciting domain of investigation. Another exciting finding has been that processing speed in infancy shows considerable cross-age stability during the first year of life. In a longitudinal study, Colombo et al. (1987) measured infants' looking time at 3, 4, 7, and 9 months. Median intermonth correlations for peak fixation duration were .34, and the median for the magnitude of habituation measured from the peak look was .40 (Colombo et al., 1987, table 5). From a study of 24 infants who participated in the VExP at both 4 and 6 months, Canfield et al. (1995) report a cross-age correlation of .81 for median RT.

Although the sample size of the present investigation is small, the Canfield et al. (1995) results suggest the possibility of finding some reliable indication of the stability of mean RT through the first year. Consequently, we computed month-to-month correlations for our global data. Figure 18 shows a scatter-plot matrix in which x-y scatter plots with regression lines are shown

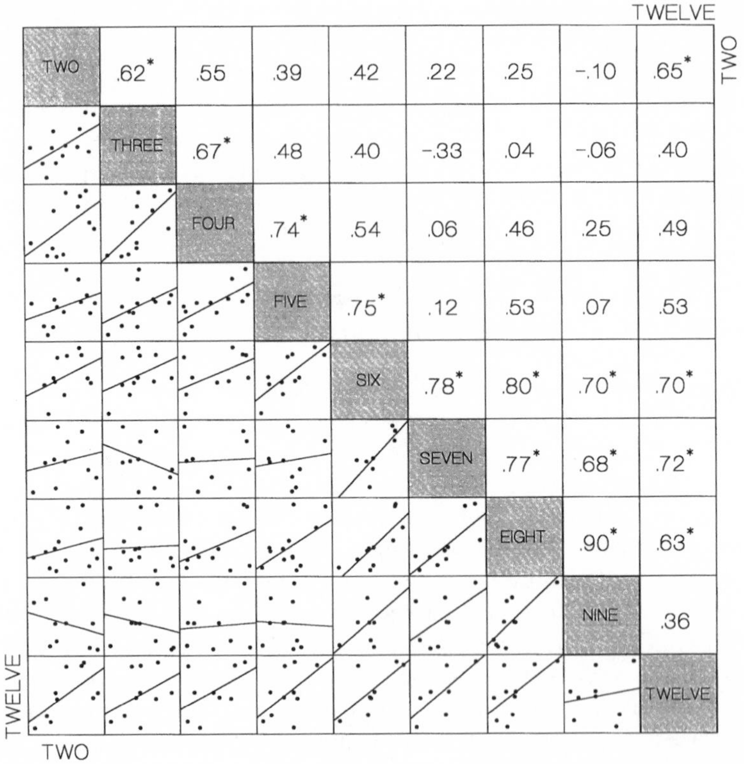

FIGURE 18.—Scatter-plot matrix showing intermonth stability of mean RT. Below the diagonal, x-y scatter plots are shown with regression lines. Above the diagonal, corresponding Pearson r's are provided. Row and column headings are provided along the shaded diagonal. * p < .05.

beneath the diagonal and Pearson r's are shown in the corresponding cells above the diagonal. Each off-diagonal cell represents either the scatter plot or the correlation between the column month and the row month.

The pattern of correlations is very systematic. The most consistent trend is that correlations between adjacent months are consistently higher than are the correlations between nonadjacent months. Every correlation between adjacent months was statistically significant (minimum r = .62, p = .025). In addition, there is a general trend of declining r's as an inverse function of both the age of the infant and the span of time between assessments. From the ages of 2–5 months, the magnitude of the correlations computed across more than a 1-month time span falls off more rapidly than for the later months. There are 22 nonadjacent intermonth r's involving the 2–5-month assessments. Of these 22 correlations, only one is statistically significant.

Therefore, we conclude that there is little evidence for stability beyond 1 month during the early months.

The pattern of intermonth correlations for the nonadjacent months from 6 to 12 months is nearly the reverse of that found for the 2–5-month period: all but one of seven correlations spanning more than 1 month are statistically significant. Furthermore, the magnitude of these coefficients is quite high (median $r = .70$). Inspection of the scatter plots provides no support for the hypothesis that these correlations are a result of one or two consistently slow or fast infants at every age. However, obvious cautions apply when results are based on such small samples. It is curious to note that only the 9–12-month correlation is not significant for the 6–12-month period. This may be a reflection of the fact that infants are achieving asymptotic performance at this time, leading to a restriction in the range of scores. Some support for this view is that the average range of scores is 189 ms at 2–8 months but that that average drops to 137 ms at 9 months and to only 84 ms at 12 months.

The general pattern of correlations found in Figure 18 suggests the possibility of different types of stability in the early months as compared to the later months. Interestingly, 6 months is the age at which the transition appears to take place—the same age at which the latency histograms changed shape to reveal clear bimodality (see Figure 6 above). In an attempt to simplify the pattern of age differences, we employed factor analysis to reduce the matrix to fewer dimensions.

Factor analysis of a correlation matrix representing the temporal patterning of a single response measure for a sample of individuals was classified by Cattell (1957, 1966) as the T-technique. This technique is known to suffer from several limitations, which Cronbach (1967) identified as arising from factoring a matrix that conforms to a simplex structure (i.e., when correlations are highest along the diagonal and fall off systematically as the duration between assessments increases). Cronbach argued that changes in factor loadings cannot be linked to specific ages because the point on the age scales will change depending both on the method of factor rotation and on which specific ages are included. However, as pointed out by Wohlwill (1973), "Neither of these criticisms appears especially relevant to the limitations of factor analysis applied to simplex-type correlation matrices, since for any factor-analytic problem the results could be expected to vary as a function of the method of rotation employed, or the particular measures (in this case, occasions) included in the matrix" (p. 270). A second point made by Cronbach, and amplified by Wohlwill, is that factors derived from a matrix with a simplex structure will be dominated by the single dimension of proximity.

Although the matrix in Table 8 shows some simplex structure, the pattern of decreasing size of the correlation with increasing temporal separation between measures is by no means uniform. The matrix is, however, domi-

TABLE 8

VARIMAX-ROTATED FACTOR LOADINGS OF INTERMONTH
CORRELATIONS OF MEAN RT

Age	Factor 1	Factor 2
2 months	.126	.776
3 months	−.124	.917
4 months	.296	.732
5 months	.517	.648
6 months	.829	.427
7 months	.904	.114
8 months	.937	.153
9 months	.905	−.026
12 months	.814	.192
Eigenvalue	4.950	1.955
Variance explained (proportion)	.47	.30

nated by the very large correlations between consecutive months. The result to be expected from factoring a highly simplex-structured matrix is a large first factor with loadings that increase monotonically over occasions and one or more additional smaller factors, also with loadings that increase monotonically. Certainly, a result that reflects nothing but temporal proximity and increasing performance over time is of little interest. However, our matrix appeared sufficiently nonsimplex that we believed that the factor analysis might reveal a more interesting pattern.

A principal components method of factor extraction was followed by varimax rotation. Two factors emerged with eigenvalues greater than 1.0. The rotated factor loadings (shown in Table 8) do not conform to the results that would be expected from factoring a simplex matrix. In fact, the loadings suggest a somewhat discontinuous developmental trajectory, which is reflected by a simple structure. The first factor (accounting for 47% of the variance) reflects RT from 6 to 12 months, and the second factor (accounting for 30% of the variance) reflects RT from 2 to 5 months. The pattern of factor loadings revealed a familiar trend: distinctly different patterns of data are seen during the 2–4-month period as compared to the 6–12-month period, and data from the 5-month assessment appear transitional. Again, this is not a pattern that simply mirrors age-related improvement in performance. Although Factor 1 shows an age-related trend toward increasing loadings over time, Factor 2 shows the reverse pattern. Importantly, both factors show a discontinuity in the loadings occurring at the 5-month assessment. This may reflect the existence of different processes influencing RT in the early and later months. It is also interesting to recall that it was not until the 6-month assessment that the fastest infants began to show asymptotic performance.

CONCLUSIONS

Considered as a whole, and taken together with research by others, our evidence strongly suggests that saccade RT represents a developmental dimension along which individual infants develop in an exponential manner. In light of research by Jacobson et al. (1992) and others, it appears that saccade RT is one of a set of observable responses that all relate to an underlying construct that may be labeled *processing speed.* However, some of our findings suggest that the nature of this dimension undergoes some kind of reorganization when infants are between 4 and 6 months of age. Consequently, the relation between saccade RT and other measures of processing speed in infancy may not be stable across development.

Our data also suggest that infants' growth curves are not identical. Rather, infants' behavioral change across age seems to take one of several forms, some of which we have observed in our sample. All these forms can be characterized as asymptotic exponential functions, albeit with different parameters. But the degree of variability among infants suggests that the development of some groups of infants may be better described using some other function (see Figure 17 above). Our data do not support the strong concept of growth that postulates that, without error variance, growth curves for all infants would be well described by a single parameterization of a particular growth function. Instead, a weaker concept of development must be applied—one that allows for substantial individual variation in shape.

Our data might be seen as posing a challenge to Kail's conclusions from studies of age changes in processing speed during childhood and adolescence. In the absence of longitudinal data, Kail cannot observe whether individual children develop in an exponential manner that conforms to the specific parameterization he found for group data. As can be seen in Figure 17 and Table 7 above, the babies in our study began their decline in RT at different starting points, developed at different rates, and maintained significant stability of individual differences across time. It is clear that the average individual function, much less a population function derived from cross-sectional data, would provide no hint about the diversity of these developmental trajectories.

It may still be the case that our data represent the development of a domain-general central mechanism that changes exponentially. But, since we used only one task, we cannot determine whether individuals have identical growth curves for other speed tasks. A demonstration that the parameters of individual infants' growth curves from another speed task (such as fixation duration) are highly similar to their growth curves for RT would provide important support for a central mechanism such as has been proposed by Kail (1993b), Hale (Hale & Jansen, 1994), and others. Nevertheless, longitudinal data collected during childhood and adolescence will be necessary to show that the exponential growth and domain-generality hypotheses apply to individual development.

When considering individual growth, the question of what constitutes similarity in the growth functions for different tasks must be confronted. Our findings suggest one possibility. That is, growth functions for different tasks within an individual might be considered similar if they both belonged to the same prototypical family of curves—as, for example, one of the three parameterizations of the exponential function that we identified (see Figure 17 above). The shape of the growth curves for different tasks should be highly similar within individuals, but the shape may vary across individuals. Burchinal and Appelbaum (1991) suggest clustering techniques that could be helpful in carrying out such an analysis.

VII. DEVELOPMENTAL FUNCTIONS AND INDIVIDUAL DIFFERENCES IN REACTION-TIME VARIABILITY AND ANTICIPATION

In this chapter, we continue the application of a developmental function approach by examining age changes and individual differences in the remaining VExP measures under investigation. We first examine our measure of reaction-time variability, the standard deviation of RT (SDRT); then we investigate our two anticipation measures, percentage anticipation (%ANT) and median anticipation latency (ANTL).

STANDARD DEVIATION OF REACTION TIME (SDRT)

The general practice of analyzing age change is nearly synonymous with the analysis of mean differences in level. However, age or group differences may be revealed, either alternatively or in addition, by differences in spread (Appelbaum & McCall, 1983; Keppel, 1982; Wohlwill, 1973). We now shift our focus to ask about the presence and nature of age-related change in the variability of RT over trials. We are not aware of any published studies using the VExP in which intertrial variability in RT is reported or discussed. The measure has, however, received some attention by researchers studying the relations between elementary cognitive processes and measures of psychometric intelligence in adults (Eysenck, 1986; Jensen, 1992b; Larson & Alderton, 1990; Vernon, 1987) and by researchers studying normal cognitive aging (Hale et al., 1988). Findings from these areas suggest the potential value of investigating the dimension of RT variability in relation to age changes and individual differences in infant performance.

For example, in the context of studies seeking relations between RT and psychometric g, Jensen (1992a) reports that individual differences in intertrial variability of RT are more predictive of individual differences in g than are differences in mean RT (Jensen, 1982). He interprets these findings as supporting Eysenck's speculations that individuals differ in the prevalence of neural transmission errors and that this is the mechanism underlying the correlation between RT and *g* (Eysenck, 1986; see also Myerson, Hale, Wagstaff, Poon, & Smith 1990). Given these empirical findings and conceptual models for relating interindividual variability to both normative aging and individual differences in cognitive ability, we believe that researchers studying infant information processing should begin examining measures of interindividual variability.

Choosing a Measure of Spread

Prior to our investigation of age-related change in variability of RT, it is necessary to choose an appropriate measure of spread. However, we must first distinguish between two basic types of variability. Following Hale et al. (1988), we distinguish *dispersion* (trial-to-trial variability in RT within a person) from *diversity* (person-to-person variability in average RT within a group). Our concern here is only with dispersion.

Our choice of a measure of spread was also based on Ratcliff's (1993) studies of simulated RT distributions. We sought a measure the estimate of which has a small standard deviation and that is relatively stable in the presence of outliers. In addition, we wanted a measure that was minimally affected by the application of RT cutoffs. For RT distributions, Ratcliff recommends using either the quartile deviation ([Quartile 1 − Quartile 3]/2) or the conventional standard deviation. Of the two, we favored the standard deviation because it is a more familiar index and will facilitate comparisons between our findings and those of other researchers. However, because the standard deviation is more sensitive to the choice of cutoff than is the quartile deviation, Ratcliff (1993) recommends that, in situations where there are differences in the means of the distributions being compared, one should also report quartile deviations to confirm the trends seen in the standard deviation. Accordingly, Table 9 includes the mean of the standard deviation and the mean of the quartile deviation averaged across infants for each month. Although we report analyses for the standard deviation, we carried out all our computations using both measures of spread. In no case were there systematic or statistically significant differences between the results obtained with the two indices.

Independence of Mean RT and SDRT

It has long been known by researchers studying response latencies that SDRT and mean RT tend to be highly correlated (Luce, 1986). Therefore, any useful analysis of SDRT depends on demonstrating that it provides some information that is independent of the mean. In the present study, we are specifically concerned with the degree of independence between the mea-

TABLE 9

Mean SDRT and Quartile Deviation as a
Function of Age for Entire Session

Age	SDRT	Quartile Deviation
2 months	157	114
	(28)	(24)
3 months	155	100
	(42)	(36)
4 months	144	95
	(39)	(34)
5 months	121	79
	(31)	(29)
6 months	111	70
	(29)	(31)
7 months	89	53
	(26)	(22)
8 months	82	51
	(22)	(20)
9 months	81	44
	(27)	(12)
12 months	83	51
	(23)	(19)

Note.—Standard deviations are given in parentheses.

sures in terms of their relation to age. If the two measures are wholly redundant, then we would have little justification for entertaining a parallel set of analyses for both. If, on the other hand, the two measures provide independent information about age changes, then the analysis of SDRT will serve to enrich our description of what develops and will eventually lead to a more complete explanation of development in this domain.

In our study, we confirmed that mean RT and SDRT are indeed highly correlated, with nearly 70% shared variance (r [117] = .833, $p < .001$). This is not an unusual finding in studies of reaction time, and it is to be expected of distributions that have a fixed origin but that are effectively unbounded at the other end (Luce, 1986; Wohlwill, 1973). Both variables are also highly correlated with the age of the infant. The correlation between age and mean RT is $-.714$ ($p < .001$), that between age and SDRT $-.678$ ($p < .001$). The problem of independent relations with age can be phrased as a semipartial correlation question (also known as a *part* correlation; Hays, 1981); namely, does SDRT have a significant linear relation to age after the linear variance that SDRT shares with mean RT has been removed? Any residual variance that is correlated with age must then be due to the unique relation between age and SDRT, independent of the mean.

Computing the semipartial correlation revealed a small but significant linear relation between age and the part of SDRT that is not correlated with mean RT ($r = -.14$, $F[1, 103] = 6.81$, $p < .025$). This suggests, not only that infants become faster as they get older, but also that their RTs become less variable from one trial to the next. Furthermore, the result suggests that additional investigation of age-related change in SDRT is warranted and may lead to a more complete explanation of the development of processing speed than would result from a singular focus on the mean RT. One interesting issue that arises when considering both mean RT and variability in RT concerns the primary site of developmental change. That is, is it possible to determine whether age-related changes in the shape of the RT distribution are due primarily to an increase in processing speed or to a decrease in response variability? We will revisit this issue after describing age changes and individual differences in SDRT.

Group and Individual Growth Curves for SDRT: Age and Sequence Type

Individual growth curves for mean global SDRT are shown in Figure 19; the group curve using the global data is shown in the top left-hand panel of Figure 19a. The figure reveals that the age-related decline is consistent and substantial, with the 2- and 3-month-olds being nearly twice as variable as the 8–12-month-olds. Little change in SDRT appears after 6 months. The individual curves reveal that age changes in SDRT are smaller and less uniform than they are for the mean RT. Also, compared to the individual curves for RT, the individual SDRT curves show less consistent indication of an underlying asymptotic exponential.

We reported a high correlation between our measures of mean level and spread, and the corresponding growth curves for the two measures are quite similar; nevertheless, they are far from fully dependent. An infant showing substantial decline in SDRT from one month to the next does not always show a decline in the mean, and, likewise, an infant showing a large increase in SDRT does not necessarily have a higher mean RT (see Figure 10 above and Figure 19). For example, infant 7's mean RT declined nearly 100 ms from the 2- to the 3-month assessment, but this was accompanied by an *increase* in SDRT from 161 to 186. Similarly, infant 14's SDRT dropped nearly 100 ms from 4 to 5 months, but mean RT was nearly identical at the two assessments.

Infants tended to show their most consistent decline in SDRT after 4 months. SDRT fell rapidly after this age for 10 of 13 infants, and eight of these were never subsequently more variable than at their 4-month assessment. Only two infants showed any sizable decline in SDRT after 8 months.

Beyond noting the presence and direction of change in these growth curves, we can describe the group function as a gradually declining linear function with the presence of a single bend. However, unlike the situation with RT, we have here little reason to pursue a more precise mathematical description of these individual curves. Our search for a precise description of age-related change in RT was guided by Kail's hypothesis of an exponential growth function for processing speed. In the case of SDRT, we have no a priori hypothesis and are aware of no growth models that have described age changes in spread for processing-speed variables. In addition, with only a single data point for each infant at each month and considerable month-to-month variation in SDRT, we could not meet the assumption of monotonicity, thus precluding the use of nonlinear functions on mathematical grounds as well (Burchinal & Appelbaum, 1991). Given these factors, we chose to limit our growth-curve analysis to polynomial models.

In addition to the previously mentioned benefits of working with polynomial models, their analytic potential is also more fully implemented in statistical computing software. Thus, it becomes possible for us to address all our questions about age-related change in SDRT, possible differences in the pattern of age change for the three sequence types, and possible individual differences in the parameters of the growth functions within the context of a single random-regressions mixed-model analysis. In the following section, we present our mixed-model analysis of group and individual age trends in SDRT; then we consider the stability of individual differences.

Table 10 shows the mean of SDRT for each age level separately for each sequence type. For six of the nine assessments, SDRT was lowest during the IR sequence. At 5 and 12 months, it was lowest during the L-R sequence and, at 4 months, lowest during the L-L-R sequence. Averaging across ages and infants, the mean SDRT was 105 during IR, 108 during L-R, and 113 during L-L-R (see Table 10).

Group and individual growth curves were explored in a random-regressions mixed-model analysis using SAS PROC MIXED. We used the REML method and selected the unstructured random-effects covariance structure option. Infant and sequence type were entered as classification effects, and fixed regressions were estimated for sequence type and for the linear and quadratic slopes across age, but separately for each sequence type. Furthermore, random regressions were computed for the intercept and linear slope parameters of individual growth functions. Note again that this is not the full polynomial model that we would ideally estimate. We believe that a more complete model should include a cubic term as well as its interactions with sequence type as fixed effects and would also enter the linear, quadratic, and cubic terms as random effects. Unfortunately, we were unable to get convergence with any model more complete than the one we present.

The fixed-effects test for sequence type revealed that SDRT does not differ across the sequences when age is ignored ($F[2, 277] = 1.12$, $p = .33$). However, the tests for linear and quadratic age trends revealed both a significant linear drop in SDRT with age ($F[3, 285] = 18.60$, $p < .0001$) and a significant quadratic slope ($F[3, 285] = 7.12$, $p < .0001$). These age trends were also seen at each level of sequence type. The linear slope was large and

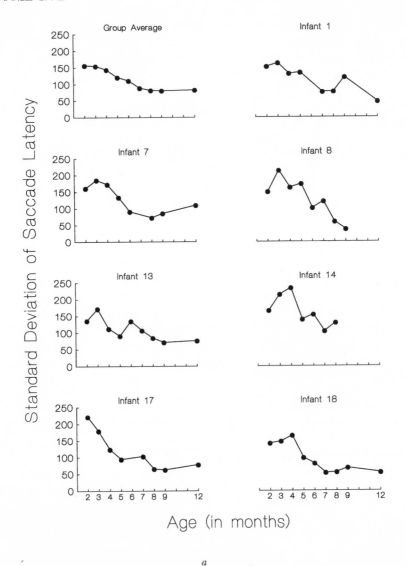

FIGURE 19.—Mean standard deviation of saccade latency (SDRT) as a function of age (2–12 months) for the group as a whole and for each infant.

significant separately for sequence IR (estimated slope = −26 ms/month, $t[285] = -4.66$, $p < .0001$), sequence L-R (estimated slope = −22 ms/ month, $t[285] = -4.10$, $p < .0001$), and sequence L-L-R (estimated slope = −25 ms/month, $t[285] = -4.21$, $p < .0001$). These parameters did not differ across conditions ($F[2, 285] = .10$).

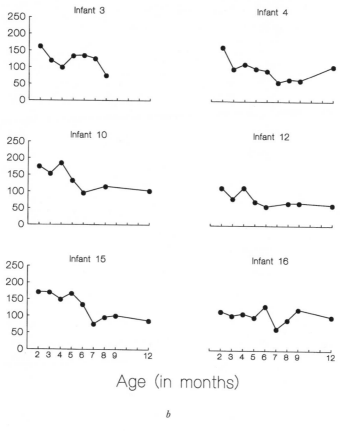

Age (in months)

b

FIGURE 19 (*Continued*)

The quadratic component of change was also significant for each sequence type individually. Sequence IR had an estimated slope of 1.25 ms/month^2 ($t[285] = 3.16$, $p < .01$), sequence L-R an estimated slope of .93 ms/month^2 ($t[285] = 2.37$, $p < .05$), and sequence L-L-R one of 1.06 ms/month^2 ($t[285] = 2.44$, $p < .05$). Again, although there were differences in the absolute magnitude of the quadratic parameter estimates, contrasts comparing slopes across sequence type did not approach significance ($F[2, 285] = .17$).

As we concluded from our analyses of mean RT, these analyses reveal considerable consistency for the three sequences. Trial-to-trial variability was nearly identical across the three sequences, as was the pattern of change. At the group level, infants showed rapid linear decline in SDRT, together with a modest quadratic trend as the rate of decline slows in the later months.

TABLE 10

MEAN (SD) STANDARD DEVIATION OF REACTION TIME
(SDRT) BY AGE AND SEQUENCE TYPE

	SEQUENCE TYPE		
AGE	IR	L-R	L-L-R
2 months	141	150	162
	(35)	(30)	(43)
3 months	145	154	166
	(42)	(50)	(42)
4 months	134	134	132
	(65)	(31)	(40)
5 months	128	110	117
	(61)	(24)	(44)
6 months	98	111	117
	(42)	(36)	(68)
7 months	76	86	85
	(37)	(24)	(43)
8 months	68	78	80
	(23)	(20)	(47)
9 months	68	75	85
	(34)	(30)	(25)
12 months	85	70	73
	(39)	(24)	(31)
Mean (SD)	105	108	113
	(32)	(33)	(35)

Individual Differences in SDRT

Finding somewhat regular age-related change in SDRT is consistent with
the interpretation that performance variability represents a developmental
dimension. This hypothesis would receive considerable additional support
were we to find evidence that individual differences in SDRT persist across
time. In the sections that follow, we explore the nature of individual differences in the general shape of the growth function and cross-age stability of
individual differences.

Individual Differences in the Shape of the Growth Function

To address the magnitude of individual differences in SDRT, we look to
the random-effects part of the mixed model presented above. Again, because
we possess data on individual growth, we can test the hypothesis that differences between individuals represent only error variance—at least in the polynomial context.

Selection from among different growth-curve models (e.g., between a
model based on the strong concept of growth and one assuming a weak con-

cept of development) is often guided by one's theory or beliefs about the individuality or commonality of patterns of growth. This situation was clearly articulated by Burchinal and Appelbaum (1991), who noted that the "selection of a growth curve model depends on what the investigator knows and believes. If the investigator believes in the individuality of growth (i.e., that nontrivial differences exist among the individuals in the patterns of growth they display), then a method that estimates individual or prototypic curves should be selected. On the other hand, belief in the universality of the developmental function corresponds to estimating group growth curves" (p. 30). The random-regressions model provides a statistical test of these assumptions. If individual growth curves are assumed to fit a polynomial of order n, one can directly test whether different parameterizations are needed for different individuals. A nonsignificant variance component associated with individuals would suggest, on purely statistical grounds, that the estimation of individual or prototypical growth curves is unwarranted in the particular sample under investigation.

The random portion of the model estimated above could potentially test for the presence of significant individual differences in three parameters estimated in the fixed portion of the model; namely, the intercept, linear slope, and quadratic slope. As noted above, the most complex model we were able successfully to estimate included only the intercept and linear slope as random effects. Nevertheless, the model provided some evidence for individual differences in the intercept ($Z = 1.92$, $p < .06$). However, evidence for differences in the linear age change was even more equivocal ($Z = 1.54$, $p = .12$).

Results from the reduced model indicate that infants may differ with respect to their initial SDRT, but they appear not to vary appreciably in the rate at which SDRT declines as a function of age. Because of the small sample size and our inability to estimate the full model, we view these findings as quite preliminary. However, they do suggest that the shape of the growth function for SDRT is similar across infants but that individual infants have different starting points.

Stability of Individual Differences

We describe our stability findings for SDRT in the scatter-plot matrix of Figure 20. The lower quadrant containing the intermonth scatter plots indicates a simplex structure. Scatter plots between consecutive months reveal substantially more stability than do the off-diagonal plots. The median correlation between consecutive months is .47, whereas the median r between months is only .30 when lagged by one and .28 when separated by two months. Unlike our findings for mean RT, few of the intermonth correlations are statistically significant. Furthermore, beyond the appearance of a simplex,

101

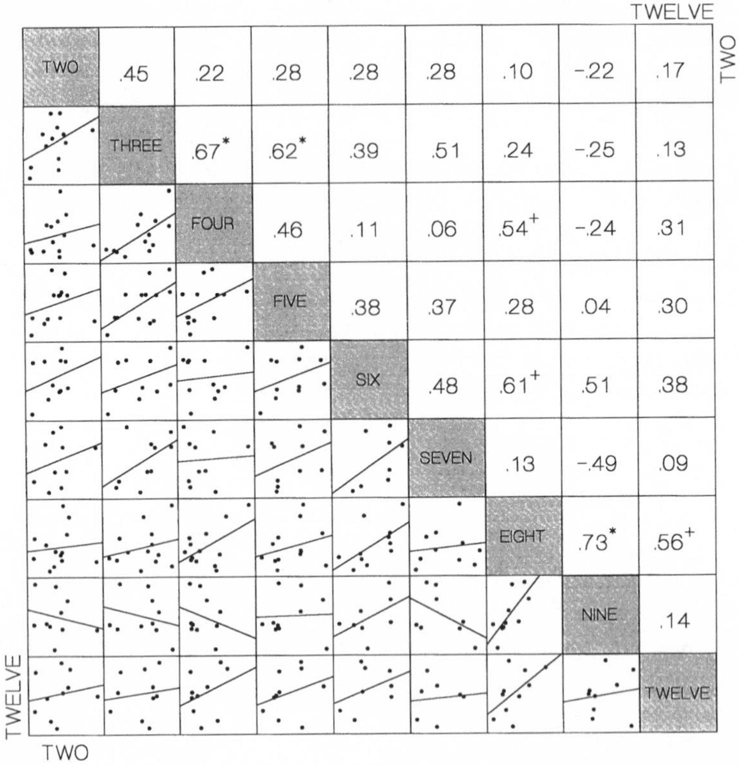

TWELVE

	TWO									TWO
TWO	.45	.22	.28	.28	.28	.10	−.22	.17		
	THREE	.67*	.62*	.39	.51	.24	−.25	.13		
		FOUR	.46	.11	.06	.54+	−.24	.31		
			FIVE	.38	.37	.28	.04	.30		
				SIX	.48	.61+	.51	.38		
					SEVEN	.13	−.49	.09		
						EIGHT	.73*	.56+		
							NINE	.14		
TWELVE									TWELVE	

TWO

FIGURE 20.—Scatter-plot matrix showing intermonth stability of mean intertrial standard deviation of RT (SDRT). Below the diagonal, x-y scatter plots are shown with regression lines. Above the diagonal, corresponding Pearson r's are provided. Row and column headings are provided along the shaded diagonal. * $p < .05$, + $p < .10$.

no general pattern emerges. Many of the correlations are, however, of a similar magnitude to other measures of infant information processing, such as fixation duration and novelty preference (Colombo & Mitchell, 1990; Colombo, Mitchell, Coldren, & Freeseman, 1991; Rose et al., 1988).

There was some indication that our small sample resulted in generally lower stability coefficients than may exist in the population. Influence plots for the intermonth correlations revealed a consistent pattern of outlier influence. In the majority of cases when an intermonth r was .30 or less, the associated influence plot revealed a single large negative outlier. Specifically, there were 20 r's that were initially at or below .30. When a single outlier was removed, 13 of those correlations rose above .30. Furthermore, influence plots from three of the remaining seven correlations had two very influential points, and the effect of removing them was to raise those r's above .30 also.

The effect of outlier removal was generally substantial, both for the 13 relations in which a single outlier was eliminated and for the three relations in which two outliers were removed. In both cases, the *r*'s were raised to a median of .45, nearly the same magnitude as the unadjusted *r*'s between consecutive months. We also examined influence plots for the intermonth correlations that were substantial before any outliers were removed. Importantly, these plots were very well behaved, indicating that the size of the initial correlations that we reported could not be attributed to one or more large positive outliers.

Conclusions

Our data cannot by themselves provide strong support for the conclusion that SDRT represents a measure of more general behavioral variability. However, they do suggest that it may be a useful measure of behavioral development that is partially distinct from mean RT. Age changes in SDRT are somewhat regular, declining in a mostly linear manner over the first year of life. Individual differences in the intercepts of the linear growth curves are large enough to suggest that babies differ in terms of their starting points, but, in our sample, they did not differ significantly in the rate at which they become less variable during the first year. The combination of different starting points but no slope differences suggests that we should find stability of individual differences in level. Although it was not strong, we found some evidence for this conclusion. We suggest that our stability analyses were compromised by the small sample size, making trends in the data severely vulnerable to the presence of outliers. Our post hoc exploration of the stability scatter plots suggests the likelihood that one would find statistically significant stability coefficients of .40 or greater in a moderately large sample of infants.

Again, taken alone, our findings did not reveal whether SDRT represents one measure from among a coherent set of observable responses that all relate to a more general construct of behavioral variability. However, we believe that our findings should encourage future investigators to include an analysis of SDRT in VExP studies and to explore potential measures of performance variability in other infant information-processing paradigms.

PERCENTAGE ANTICIPATION (%ANT)

The original inspiration guiding the development of the visual expectation paradigm was Haith's conviction that, relative to their importance in cognition generally, future-oriented cognitive processes had been neglected in the study of infant perceptual-cognitive development. The measurement of anticipatory eye saccades was developed as the principle index of a more general expectancy about where and when events will occur in the baby's visual world. We have already seen that nonanticipatory saccades provide a consistent marker of both age-related change and individual differences in infant information processing. Now we turn to a description of age-related changes in the frequency of anticipatory fixation shifts.

Unlike the simple stimulus-elicited responses that are aggregated to compute mean RT and SDRT, anticipatory fixation shifts appear to require more in the way of higher cognitive processes (Haith et al., 1993). Reactions are elicited, but anticipatory saccades involve some ability to mediate cross-temporal contingencies and some degree of choice. For the study of the cognitive processes in infants that involve representation, simple planning, and knowledge-informed choice, the expectancy-driven anticipatory saccade would appear to be a clear window to the infant's mind.

In part because of the presumed complexity of the underlying cognitive operations, the anticipation measure is more difficult to interpret than is RT. Assuming that infants across the 2–12-month period were equally motivated to anticipate, one would predict that mental maturity would be associated with a greater tendency to do so. However, recall that, in our analysis of existing anticipation data in Chapter IV, we found no evidence for a general age-related increase in the prevalence of anticipations during the 2–8-month period. An important confounding factor for that analysis was the wide variation in stimulus sequence complexity across studies and ages. Moreover, data from young infants have generally been collected using one set of methods, but other methods have been used with older infants. The present study represents an attempt to describe age trends in anticipation by using identical procedures, stimuli, patterns of sequential complexity, and timing for all infants at all ages. Obviously, we must be careful not to interpret behavioral differences as if they necessarily represent ability differences. Differential motivation at different ages is a potential confounding factor.

Our analysis of percentage anticipation will continue the descriptive theme already established. The first step will be to describe at the group level the possible differences in %ANT as a function of age and sequence type. For this variable we have good reason to suspect that developmental trajectories might vary as a function of the spatiotemporal predictability of the sequence. Previous research has documented that, under some conditions, babies as young as 2 months readily make anticipatory fixation shifts during a 1-second interstimulus interval (Canfield & Haith, 1991; Smith & Canfield, 1996; Wentworth & Haith, 1992). However, the ability to form expectations for regular side shifts could be sensitive to developmental change. We expect to find that all infants engage in anticipatory shifting during the first irregular portion of the stimulus sequence, regardless of age. But, as predicted from previous research, we expect that, when the sequence shifts to the highly

TABLE 11

Mean (SD) Percentage Anticipation by Age and Sequence Type

		Sequence Type		
Age	Global	IR	L-R	L-L-R
2 months	11	8	12	13
	(7)	(7)	(7)	(12)
3 months	18	10	20	20
	(12)	(10)	(13)	(18)
4 months	13	9	16	11
	(7)	(8)	(13)	(9)
5 months	17	16	14	20
	(10)	(15)	(11)	(15)
6 months	19	7	19	24
	(9)	(10)	(10)	(16)
7 months	17	10	15	20
	(10)	(12)	(10)	(12)
8 months	17	10	20	17
	(10)	(9)	(12)	(10)
9 months	18	9	24	20
	(13)	(10)	(19)	(14)
12 months	11	8	13	9
	(8)	(12)	(11)	(11)
Mean (SD)	16	10	17	17
	(3)	(3)	(4)	(5)

predictable L-R alternation, only the older infants will show a higher proportion of anticipatory fixation shifts. Similar to our analysis of SDRT, these questions as well as questions about individual differences will be explored in the context of a random-regressions mixed model.

Trends in %ANT by Sequence Type and Age

Mean %ANT for each age level, computed both with and without respect to sequence type, is shown in Table 11. Note that, when sequence type is ignored, %ANT includes both correct and incorrect anticipations (e.g., as in the IR and L-L-R sequences when the infant shifts to the right side after only one left-side stimulus has appeared). Table 11 shows that the range of mean %ANT across all sequence types is quite small, with a low of 7% and a high of 24%. No obvious age trend is apparent when %ANT is computed globally or separately for the IR sequence. The two predictable sequences show a slight tendency toward higher %ANT for the later months, yet all sequences also show nearly their lowest %ANT at the 12-month assessment. Furthermore, for both predictable sequences, average %ANT at 3 months (20%) nearly equals the maximum rate for any age.

Again, sequence differences and group and individual growth curves were explored using a random-regressions model. As with the model exploring SDRT, we used the REML method with an unstructured random-effects covariance matrix using SAS PROC MIXED. Infant and sequence type were entered as classification effects. As before, we were unable to estimate our desired model, but we succeeded in estimating a model that included fixed regressions for the linear and quadratic components of age, both separately for each sequence type and overall. Furthermore, random regressions were computed for the intercept and both the linear and the quadratic slope parameters corresponding to individual growth functions (again, the model specified subject = individual).

Results from the fixed-effects portion of the model revealed significant mean differences in %ANT across sequences ($F[2, 262] = 12.00$, $p < .0001$). Pairwise differences between individual sequence means were significant for sequences IR and L-R ($t[262] = -3.95$, $p < .001$), and for IR and L-L-R ($t[262] = -4.46$, $p < .001$), but not for L-R and L-L-R ($t[262] = -.58$, all p values Bonferroni adjusted; see Table 11). In addition, both the linear and the quadratic slopes were marginally significant when averaged across sequence type ($F[1, 262] = 4.90$, $p < .028$, and $F[2, 262] = 5.40$, $p < .021$, respectively).

The pattern of age-related change differed somewhat depending on the sequence. For sequence IR, there was no evidence for significant linear or quadratic change (estimate = 1.5%/month, $t[262] = .78$, and estimate = $-.116\%$/month^2, $t[262] = -.85$, respectively). However, there were marginally significant trends in sequence L-R (estimate = 3.31%/month, $t[262] = 1.73$, $p < .09$, and estimate = $-.226\%$/month^2, $t[262] = -1.66$, $p < .10$, for linear and quadratic slopes, respectively) and significant trends for sequence L-L-R (estimate = 4.76%/month, $t[262] = 2.36$, $p < .02$, and estimate = $-.364\%$/month^2, $t[262] = -2.48$, $p < .02$, for linear and quadratic slopes, respectively).

Regardless of these seemingly different patterns, there was no evidence for a statistically significant difference between the sequences in the pattern of change over age. This was true for both the linear and the quadratic slopes ($F[2, 262] = .99$, and $F[2, 262] = 1.05$, respectively).

The significant positive linear slopes suggest increasing %ANT as a function of age, but, when coupled with a negative quadratic slope, the curve becomes an inverted U-shaped function, indicating that %ANT increases and then decreases during the 2–12-month period. It is easy to see what is the cause of this inverted U: it is the sharp decline in %ANT between the 9- and the 12-month assessments. This decline is seen in each of the three sequences, and its saltatory nature may indicate the presence of a qualitative change in the developmental function. That is, it may indicate that %ANT measures somewhat different capacities or motivations at 12 months than it does during the 2–9-month period. After looking at the data, we suspected that the overall description of development may be quite different if we excluded the 12-month data from our analyses. Therefore, we ran a post hoc analysis identical to the previous mixed-model analysis but included data only for 2–9 months. As for the previous analysis, we were interested in the linear and quadratic age trends and their possible differences across sequence type.

When the 12-month data are removed, the overall age trends are radically altered. For no sequence does the linear or quadratic trend on age approach significance (all p's $> .15$). Inspection of individual parameter estimates shows, however, that for sequence L-R the estimate for the linear slope is changed from positive to negative and the quadratic slope changed from negative to positive. Now the estimated curve is inflected in the opposite direction, becoming a faster-than-linear increase in %ANT for the 2–9-month period. The only other evidence for an age-related increase in %ANT came from a model in which only the linear slope was estimated for each sequence. When the quadratic term was removed, a marginally significant positive linear slope was estimated for sequence L-R alone (estimate $= 1.16$, $t[250] = 1.74$, $p < .09$).

These findings suggest only weak evidence for systematic age-related change in %ANT. Although the linear slopes were positive and significant when estimated from the 2–12-month data using a model that included a quadratic term, it appears that this was due more to the presence of the quadratic than to any real increase in %ANT. Because the 12-month %ANT was so small and the change from 9 months so large, any quadratic function fit to that pattern of data would by necessity need to increase rapidly before decreasing to fit the 12-month values. Thus, as a result of the mathematical model used, the size of the linear parameter was artificially inflated. This interpretation is bolstered by our finding from the analysis of 2–9-month data. When the quadratic term was removed from the model, the size of the linear parameters was greatly reduced, leaving only the L-R sequence showing a slight linear increase in %ANT as a function of age.

The most consistent finding with %ANT is more anticipation for the two predictable sequences (which also occur later in the sequence) as compared to the initial IR sequence and somewhat greater anticipation for the later ages for the L-R sequence, except at 12 months. It is striking to note that, similar to the findings from previously published studies that were described in Chapter II, age trends in %ANT are quite weak. However, because we found a marginally significant linear age trend only for the L-R sequence, it is possible that our use of multiple sequences may have obscured more consistent age trends than might be observed using a single, alternating stimulus sequence. Because our age trends were most lawful for the L-R sequence and only for the 2–9-month assessments, our analysis of differences among individual growth curves will be carried out using only those data.

107

Individual Differences in %ANT

Individual Differences in the Shape of the Growth Function

The slight linear trend of the group growth curve may result from many patterns of individual growth. For example, one subgroup of infants may show a large increase in %ANT with increasing age, while another may show either no increase or even a decrease. Individual growth curves from 2 to 9 months for %ANT in the L-R sequence are shown in Figure 21, along with the group curve. The individual curves of the majority of infants reveal a mostly unsystematic pattern of age change. It appears that either a very high or a very low rate of anticipation may be seen at any age in a particular infant's data. However, several infants show a somewhat steady linear increase (infants 1, 7, 10, 12, and 17), and others show either no increase or large month-to-month fluctuations.

It would appear problematic to interpret the systematic nature of the age-related increase in these five infants as evidence for an increase in their *ability* to anticipate. Note that the maximum anticipation rate for these infants is similar in magnitude to that for several other infants who had achieved a rate of 45%–55% anticipation sometime in the first 6 months of life. Thus, some infants appear capable of anticipating at a high rate as early as 2 or 3 months of age, but they rarely maintain a high anticipation rate for consecutive months and may anticipate at a much lower rate at later ages (e.g., see infants 3, 4, and 14 in Figure 21). This pattern suggests a more central role for motivation and choice in the anticipation measure when compared to reaction time.

As for the analysis of SDRT, we used the variance estimates for the intercept and linear slope from the random-regressions model to assess the significance of individual differences in the shape of the growth function. We chose to limit our analysis to only those data that appear to be at least somewhat lawfully related to age—data from the L-R sequence for the 2–9-month assessments. Furthermore, we have no a priori reason for investigating more complex models than the linear. Even though quadratic, cubic, or higher-order polynomial trends in the individual growth data may exist, we felt that our %ANT data were better suited to addressing simple questions about the existence of individual differences in the intercepts and slopes of the linear growth functions. And, as we have already seen, the inclusion of the quadratic parameter in the model can easily distort the interpretation of the linear slope. Therefore, we asked if individual babies differ in their initial rate of anticipation and in their rate of change in anticipation during the 2–9-month period.

Computation of a random-regressions model provided no support for the hypothesis that individuals differ significantly in the intercept ($Z = 1.16$,

$p > .20$) or the slope ($Z = 1.54$, $p > .10$) of the linear growth curve for %ANT. Given the substantial differences between curves for particular individuals as shown in Figure 21, it appears that one reason for the nonsignificant findings may be the high level of month-to-month variation in %ANT within a given infant's data, coupled with a lack of power.

We suggest that the within-infant variability represents the normal day-to-day fluctuations in infant performance that plague infant research generally (Fischer & Hogan, 1989). It is likely that a more lawful growth curve would be observed were one able to measure optimal performance at each assessment (Fischer, 1980; Fischer & Hogan, 1989). Future research will need to focus on reducing the within-subject month-to-month variability in anticipation performance, motivating infants to achieve optimal performance, and possibly making the task more challenging to amplify age trends. Judged in terms of the present findings, the hypothesis that %ANT is a viable measure of some underlying dimension of behavior, and that individual infants develop in a lawful manner along that dimension, appears to be in jeopardy.

Stability of Individual Differences

We first looked at stability of individual differences in %ANT separately by sequence type, then considered stability for correct anticipations by themselves. The most lawful findings were obtained for data within the L-R sequence alone or for the correct anticipations in the combined L-R and L-L-R sequences.

Intermonth correlations indicated that individual differences in %ANT are not stable for sequence IR. Of the 36 intermonth correlations, only 15 are positive, and only one is marginally significant. No discernible pattern emerges from this analysis, leading to a conclusion of no month-to-month stability.

Evidence for intermonth stability for %ANT in the L-R sequence is also quite weak, although there are some indications of stability after 6 months. The scatter-plot matrix of intermonth correlations for sequence L-R is shown in Figure 22. Eleven of 36 coefficients are negative, and all but one of these occurs within the first 5 months. The median intermonth r for all pairs of assessments before 6 months is .09, whereas the comparable median r from 6 to 12 months is .49.

As was the case in our stability analysis of SDRT, many of the negative and small positive r's arose as a consequence of one or two very influential points—and this was especially true for intermonth correlations beginning with the 6-month assessment. For the 10 post-5-month scatter plots with a single large negative outlier, removing that one influential score raised the median intermonth correlation from .49 to .71. In the same way, removal of

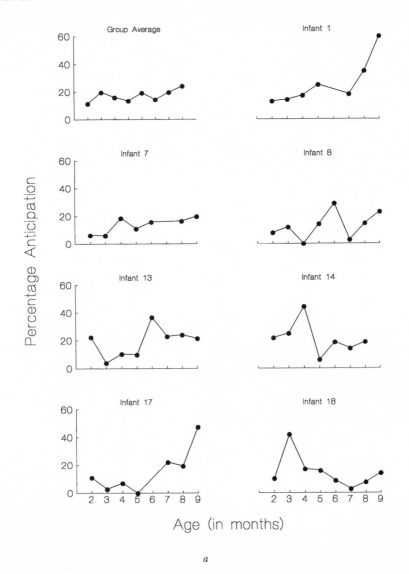

Percentage Anticipation

Age (in months)

a

FIGURE 21.—Mean percentage anticipation (%ANT) in the L-R sequence as a function of age (2–9 months) for the group as a whole and for each infant.

outliers for %ANT when L-R and L-L-R were combined caused the median intermonth correlation after 6 months to increase from .53 to .58.

Our findings confirm previous research (Canfield et al., 1995) indicating no stability of individual differences from 4 to 6 months of age in %ANT for an L-R sequence. However, there appears to be some stability in measures

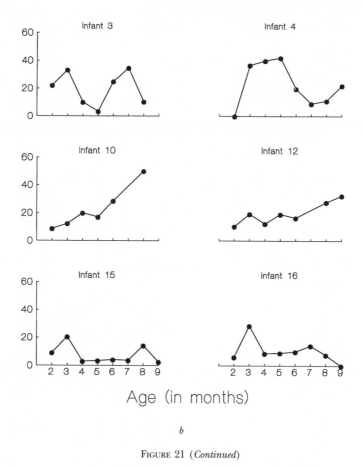

b

FIGURE 21 (*Continued*)

taken from the predictable sequences during the second half of the first year. From our initial analyses, there was not much reason to suspect stable individual differences in %ANT; however, we believe that outliers played a significant role in attenuating stability that we suspect exists in the second 6 months of life. Stability of %ANT is consistent with the pattern of generally greater stability during the latter half of the first year, as suggested by our factor analysis of RT stability data presented in Chapter VI, and stronger stability of SDRT from 6 to 12 months, described earlier in this chapter.

Conclusions

Although we found fewer significant age changes and individual differences for %ANT than for RT and SDRT, some interesting trends were re-

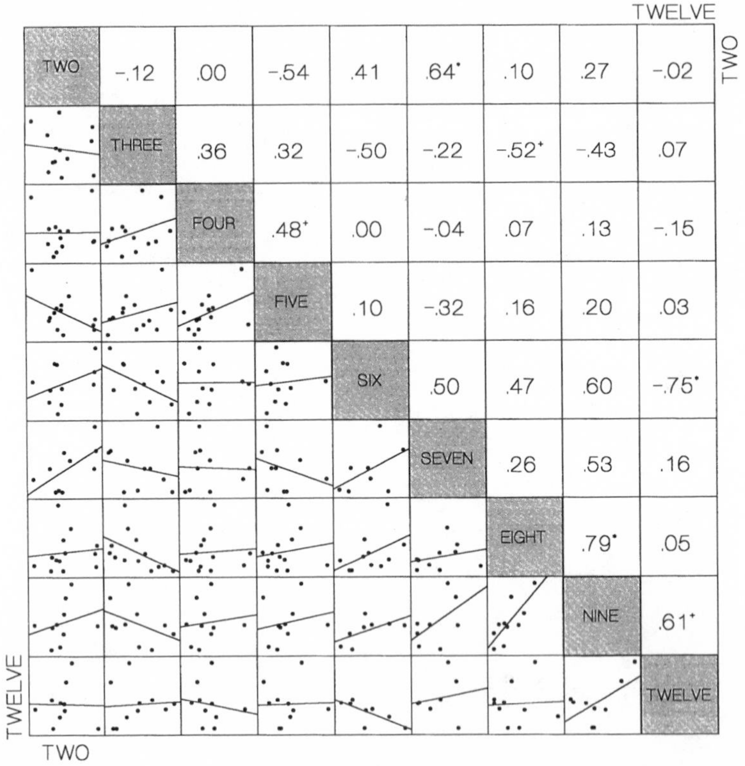

FIGURE 22.—Scatter-plot matrix showing intermonth stability of mean percentage anticipation (%ANT) in the L-R sequence only. Below the diagonal, *x-y* scatter plots are shown with regression lines. Above the diagonal, corresponding Pearson *r*'s are provided. Row and column headings are provided along the shaded diagonal. * $p < .05$, + $p < .10$.

vealed. First, there was some indication that, in terms of anticipation, VExP may be a qualitatively different task for 12-month-olds than it is for 2–9-month-olds. Almost certainly, this qualitative difference involves a change in the motivational and choice-related aspects of the sequences and stimuli that we showed the infants. Second, we found consistently greater %ANT between the predictable sequences, L-R and L-L-R, and the unpredictable IR sequence across all ages. Given that sequence type was confounded with order of presentation, it is difficult to determine whether these findings are the result of sequence predictability or some sort of warm-up effect. Finally, our analysis of %ANT suggests that stable individual differences may be negligible until the second 6 months of life.

ANTICIPATION LATENCY (ANTL)

The final variable that we investigate is median anticipation latency (ANTL). This variable is thought to reflect the time it takes for the infant to access an expectancy and initiate an anticipatory saccade to the alternate side following the offset of a stimulus. To our knowledge, this variable has appeared in the literature only once. In the Canfield and Smith (1996) study of sequential enumeration, anticipation latencies were analyzed to evaluate a hypothesis about the nature of infants' numerical representations.

ANTL is calculated using the onset time of the stimulus that was anticipated as the reference point. If an anticipatory shift precedes the onset of the stimulus, ANTL takes on negative values; if the shift follows the stimulus (by up to 100 ms), ANTL is positive. Because we did not want a single extremely early or late anticipation to be overly influential in the calculation of an average latency, we chose to compute the median anticipation latency as our measure of central tendency. Unlike the RT distributions, the anticipation latency distributions were not positively skewed, and it was common for infants to have a small number of anticipations. Under these conditions, the sample median would not be expected to give a biased estimate of the population median, as it does with RT, but would result in a measure of central tendency that would not be overly influenced by extreme scores.

Trends in ANTL by Sequence Type and Age

Means of median anticipation latencies were computed for the global data, separately by sequence and averaged across infants (see Table 12). We noted that these data were extremely variable, with the standard deviations frequently exceeding the mean value. This was especially true for the IR data (6 of 9 months) but less true for the L-R and L-L-R data (2 of 9 months and 3 of 9 months, respectively). The relatively higher variability for the IR sequence may reflect the facts that it occurred at the beginning of each session and that the stimuli appeared in a nonpredictable manner. This within-sequence variability made our models that compared %ANT across sequences mostly uninformative. Thus, in an attempt to maximize our chances of finding some patterns of age-related change, we averaged ANTL for sequences L-R and L-L-R and carried out our descriptive and inferential analyses on these data.

Anticipation latencies were combined for the L-R and L-L-R sequences to produce a single growth curve for each infant. Figure 23 shows the group growth curve for these combined data in the upper left-hand panel (Figure 23a) and the corresponding individual curves in subsequent panels. The

113

TABLE 12

Mean (SD) of Median Anticipation Latency (ANTL) as a Function
of Age and Sequence Type

		SEQUENCE TYPE		
AGE	GLOBAL	IR	L-R	L-L-R
2 months	−110	−176	−122	−110
	(147)	(199)	(145)	(142)
3 months	−165	−127	−195	−163
	(141)	(181)	(239)	(189)
4 months	−146	−115	−163	−158
	(71)	(127)	(134)	(199)
5 months	−190	−185	−186	−196
	(100)	(112)	(88)	(174)
6 months	−224	−137	−146	−298
	(87)	(186)	(127)	(141)
7 months	−298	−242	−253	−335
	(168)	(93)	(198)	(190)
8 months	−278	−122	−273	−328
	(117)	(126)	(137)	(148)
9 months	−383	−303	−368	−442
	(151)	(191)	(156)	(108)
12 months	−258	−200	−325	−253
	(158)	(243)	(115)	(156)
Mean (SD)	−228	−179	−226	−254
	(85)	(63)	(84)	(107)

group curve shows a regular linear decline in ANTL from 2 to 9 months and
then an increase from 9 to 12 months. The individual growth curves also
showed a decline for all infants from 2 to 9 months, but six of the eight
infants for whom there were 12-month data showed an increase in ANTL at
12 months.

As before, we began our modeling of group and individual growth curves
using a random-regressions mixed model. Infant was entered as a classification effect, and fixed regressions were estimated for the linear and the quadratic slopes across age. Furthermore, random regressions were computed
for the intercept and for the linear and quadratic slope parameters of individual growth functions.

Unlike the previous variables studied, these data, when analyzed, revealed no evidence of a significant quadratic trend, at either the group or
the individual level. The overall test for the quadratic was not significant ($F[1,
167] = .76$). Although visual inspection of the individual growth curves sug-
gests some quadratic or higher polynomial component, we were unable to
confirm the presence of any function more complex than the linear. We then
estimated a model in which the linear term was entered as both fixed and
random. In the fixed part of the model, the parameter estimate for the linear

trend was marginally significant ($F[1, 12] = 4.03$, $p < .07$), indicating the tendency for ANTL at the group level to decline during the 2–12-month period.

Individual Differences in ANTL

Individual Differences in the Shape of the Growth Function

We again used a random-regressions model to assess the degree to which individuals differ in the parameters of their growth functions, and we consider only the 2–9-month data. Since the fixed-effects trend analysis indicated nothing more complex than a simple linear trend on age, our model estimated variances for only the intercept and slope of the linear function.

The random part of the model provides no evidence to suggest that individuals differ in either the intercept ($Z = .78$) or the slope ($Z = .98$) of the linear growth function for ANTL. To a large extent, the estimated individual linear functions remained parallel. Although our findings with RT showed a general decrease in the magnitude of individual differences across age, there is some suggestion that individual infants differ more in ANTL as a function of age. The y intercept for ANTL has a range of only 95 ms, but the range increases (beginning at about 6 months) to nearly 200 ms by 9 months. Probably the most noteworthy feature of these data is the extreme degree of variability both within infants across months and across infants at a given month.

Stability of Individual Differences

Although the shape of the growth function shows little variance between individuals, these functions may represent a stable characteristic of infants that endures through time. The intermonth stability coefficients for ANTL are shown in the scatter-plot matrix in Figure 24.

Initial analyses of intermonth correlations suggest an almost total lack of stability. Of the 36 coefficients, 14 are negative, and only one reaches statistical significance. Outliers had no discernible effect on the magnitude of the intermonth correlations before 7 months, for which the most influential values were as likely to be responsible for negative as they were for positive correlations. In contrast, for the six intermonth correlations beginning at 7 months, a single, very discrepant outlier was responsible for the three smallest r's, and there were no highly influential outliers contributing to the other intermonth r's. When the outlier is removed, the lowest intermonth correlation between 7 and 12 months is raised from $-.24$ to $.47$, and the median correlation is raised from $.35$ to $.61$. These findings parallel our previous stability findings because they indicate (1) that, as currently measured, ANTL

115

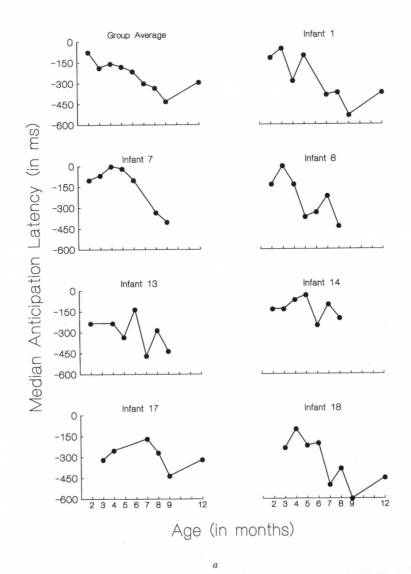

a

FIGURE 23.—Median anticipation latency (ANTL) as a function of age (2–12 months) for the group as a whole and for each infant. ANTL is averaged across sequences L-R and L-L-R.

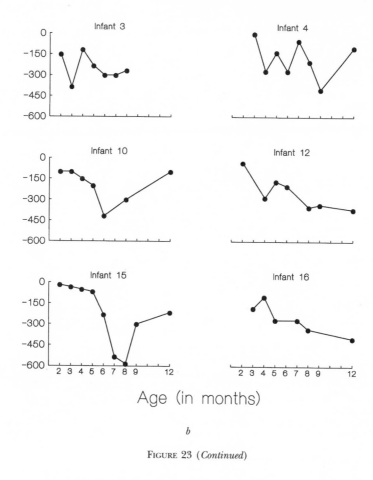

Age (in months)

b

FIGURE 23 (*Continued*)

shows no stability during the first 6 months of life and (2) that there is some indication that stability in ANTL may emerge during the second half year.

Conclusions

Our findings from the analysis of anticipation latencies are similar to those for %ANT in that they are all consistent with the hypothesis that anticipation represents a somewhat different activity for 12-month-old infants than for 2–9-month-olds. Similarly, there is little evidence of stability of individual differences in either measure, especially in the first 6 months. Finally, like the %ANT findings, ANTL showed the possibility of slight age-related change. The group average growth curve was fairly smooth, and most infants

117

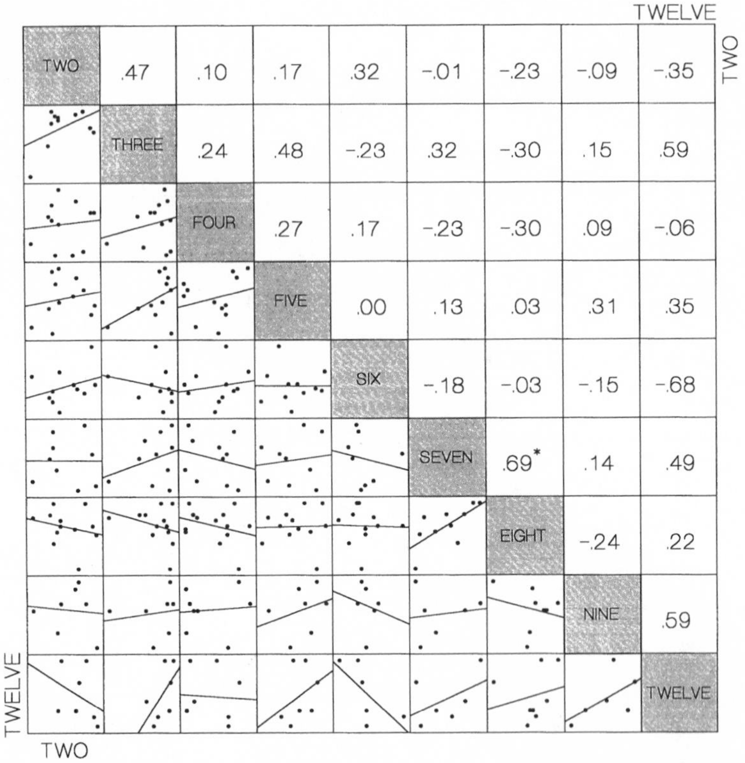

TWELVE

	TWO	THREE	FOUR	FIVE	SIX	SEVEN	EIGHT	NINE	TWELVE
TWO		.47	.10	.17	.32	-.01	-.23	-.09	-.35
THREE			.24	.48	-.23	.32	-.30	.15	.59
FOUR				.27	.17	-.23	-.30	.09	-.06
FIVE					.00	.13	.03	.31	.35
SIX						-.18	-.03	-.15	-.68
SEVEN							.69*	.14	.49
EIGHT								-.24	.22
NINE									.59
TWELVE									

TWELVE

TWO

FIGURE 24.—Scatter-plot matrix showing intermonth stability of median anticipation latency (ANTL). Below the diagonal, *x-y* scatter plots are shown with regression lines. Above the diagonal, corresponding Pearson *r*'s are provided. Row and column headings are provided along the shaded diagonal. * $p < .05$.

showed regular age-related decline until 8 or 9 months. Thus, ANTL may represent a viable measure of development during early infancy.

The most obvious interpretation of ANTL is speed of processing, but, as in the case of %ANT, it would also appear to measure some aspect of motivation to respond quickly. The higher level of choice involved in the timing of the anticipation response as compared to the reaction probably makes it more sensitive to day-to-day and moment-to-moment fluctuations in arousal, attention, and motivation. This may make ANTL a relatively poor measure of information processing, but, as with %ANT, it may also mean that it measures a more complex and therefore variable cognitive process than RT. Because of its easy availability when anticipation data are gathered, we believe that it warrants further investigation.

GENERAL CONCLUSIONS

Analyses of SDRT, %ANT, and ANTL in this chapter indicate that these measures are not as robust as RT when one is seeking to describe how they change as a function of age during the first year of life. While each measure suggested some age-related change, the developmental attributes of the putative constructs that the variables represent appear to be more subtle than are those of RT.

Prior to this study, SDRT as a measure of trial-to-trial variability had not been described in relation to changes with age during infancy. The present investigation has made several noteworthy findings concerning SDRT. For example, despite a high correlation with mean RT, a small but significant portion of the age-related variance that can be attributed to SDRT is independent of RT. In addition, SDRT decreases as a function of age. Infants may differ in their initial SDRT, but it does not appear that they differ in their linear rate of decline with age. There also appears to be little stability of individual differences, although the cross-age correlations for the later months increased substantially when outlying data points were removed.

The two measures of anticipation that we studied showed several common characteristics. Analyses of %ANT and ANTL suggest that, by 12 months, infants' motivational requirements in the VExP may have changed. We often found that, at 12 months, infants would anticipate less often and later in the anticipation interval than they did at 8 or 9 months. These findings suggest that, when attempting to develop equally motivating stimuli and settings across age groups, changes may need to be made when testing infants older than 9 months. Investigation of %ANT and ANTL also revealed that infants' growth curves did not vary significantly in terms of the y intercept or the rate of change with age. This finding is consistent with the belief that group anticipation curves are representative of all individuals within the group. A third common characteristic is that both %ANT and ANTL showed linear age effects specific to the predictable L-R and L-L-R sequences and not to the unpredictable IR sequence. From 2 to 9 months, %ANT in the L-R sequence increased slightly. During this same interval, median ANTL decreased more, that is, was faster, in the predictable sequences than in the unpredictable one. Finally, in the area of stability of individual differences, neither %ANT nor ANTL showed significant stability, although, after outliers were removed, there was some indication of stability for both measures during the second 6 months of life.

VIII. CONCLUSIONS AND FUTURE DIRECTIONS

The research reported in this *Monograph* is novel in several respects. It is (1) the first long-term longitudinal investigation of infant information processing using the Visual Expectation Paradigm (VExP); (2) the first investigation of saccade reaction time and anticipation with infants older than 8 months; (3) the first study to use converging behavioral indices to decisively separate reactive from anticipatory saccades in infants, thus confirming the conceptual dichotomy that forms the basis for the VExP; (4) the first to report findings on the standard deviation of RT in babies; and (5) the first to use a developmental function approach to investigate normative development and individual differences in the VExP.

In this chapter, we first consider the implications of our research for the interpretation of previous findings from research using the VExP. Following this, we discuss the implications for understanding continuities and change in infant information processing. In this context, we note convergences and divergences between our longitudinal findings with infants and the cross-sectional research used to test broad hypotheses about the nature of cognitive development, namely, the global trend hypothesis and the exponential growth hypothesis. Next, we use a variable criterion performance model—originally proposed to explain within-subject and between-subjects variability in RT in an adult population—to address the question of what develops in processing speed during the first year of life. Finally, we suggest several avenues for future research on infant information processing using the VExP and related paradigms.

SHORTENING THE MINIMUM RT: DOES IT MATTER?

What may be viewed as the most practically important finding from our investigation is that the previous definition of minimum RT being greater than 200 ms (our 233-ms bin), which has been used in all previous VExP studies, is likely to be incorrect. Research based on an incorrect minimum would, of course, yield imprecise estimates of average RT and %ANT. Were a large enough discrepancy found between the reported and the true averages, then formerly accepted conclusions about infant cognition based on findings from the VExP may require reevaluation. We must also entertain the hypothesis that the minimum RT that we found is somehow unique to our particular stimuli and experimental procedures. However, the similarity of our methods to those used in other studies and the uniformity of our findings across individuals and ages suggest that the 133-ms value is not far from what other investigators would find were they to submit their data to an analysis of corrective saccades and anticipation errors.

Nevertheless, the practical significance of our findings remains to be demonstrated. We are in a unique position to explore the effect of the minimum RT value because we can compare our findings that were based on the 133-ms minimum to what they would have been had we used the 233-ms minimum. However, rather than carry out a detailed set of analyses such as we reported in the previous chapters, we think it sufficient to present several key comparisons that are most likely to reveal what types of previous research may be particularly vulnerable to the effect of overestimating minimum RT.

In our data set, shortening the minimum RT leads to the reclassification of between 5% and 16% of the data ($M = 9\%$), depending on age. One way to assess the effects of this action is to compare the group average growth curves for global mean RT and for %ANT, using both minimum RT values. These plots will reveal the average difference between the means and also show how those differences vary as a function of age. The dashed lines in the upper portion of Figure 25 show the group average growth curves for mean RT. The upper line represents the growth curve obtained using the 233-ms minimum value; the lower line represents the growth curve obtained by using the 133-ms minimum. The solid line at the bottom of the figure shows the average difference between the two upper curves, that is, the amount by which the group mean RT is overestimated by using the higher minimum. This line shows that, compared to the absolute level of the means, the difference is quite small—averaging only about 15–20 ms across all ages. Although decreasing the minimum RT to 133 ms lowers the means slightly, there seems to be little or no effect on the overall shape of the age trends, as represented by the group average mean RTs. This finding reflects the fact that the number of latencies in the 133-, 167-, and 200-ms bins is small relative to the number of latencies that are included in the calculation of the mean. Therefore, we would expect that the results of our growth-curve modeling or individual differences for RT would not have been altered had we used the higher minimum.

A quite different conclusion emerges from our investigation of the effects of the minimum RT value on the calculation of %ANT and its age-related change. The dashed lines in the upper portion of Figure 26 show the

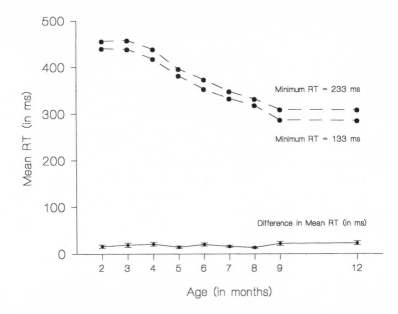

FIGURE 25.—Mean global RT as a function of age for two minimum RT values (upper lines), and difference in mean global RT ([RT using 233-ms minimum] − [RT using 133-ms minimum]) as a function of age (lower line, 2–12 months). Bars represent the standard error of the mean.

group average growth curves for %ANT. The upper line is based on the 233-ms minimum value, and the lower dashed line is based on the 133-ms minimum. In contrast to what we observed for mean RT, the relatively small number of anticipations entering into the %ANT calculation, coupled with the relatively large number of RTs in the 133-, 167-, and 200-ms bins at the later ages, results in a substantial difference in both the levels and the shapes of the growth functions for mean %ANT. The solid line in Figure 26 represents the magnitude of the difference between the dashed lines.

What this figure reveals is that the mean %ANT is overestimated when the 233-ms minimum is used and that it is overestimated by a much greater amount at the older ages. This difference is so substantial that, when 200-ms latencies are included in its calculation, %ANT at 12 months is more than double what our research suggests the actual value should be. This difference would have substantially altered our conclusions about the nature of age changes in anticipation. Not only would we have reported a much greater age-related increase in %ANT, but we would also have been less likely to note the somewhat discontinuous nature of the development of anticipation between 9 and 12 months—as indicated by our finding of a precipitous drop in %ANT and a decrease in anticipation latency. Thus, an accurate estimate

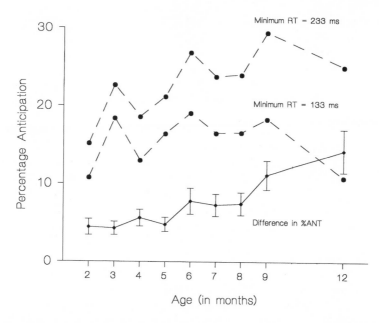

FIGURE 26.—Mean global percentage anticipation (%ANT) as a function of age for two minimum RT values (upper lines), and difference in mean global %ANT ([%ANT using 233-ms minimum] − [%ANT using 133-ms minimum]) as a function of age (lower line, 2–12 months). Bars represent the standard error of the mean.

of minimum RT appears essential for an accurate estimation of %ANT, whereas the calculation of average RT (when that average is based on a sizable sample) is not likely to be greatly affected. Moreover, the importance of one's choice of the minimum RT increases with increasing age.

Reevaluating Previous Research Findings

We think that it is important to assess the potential effects of the past practice of classifying all latencies less than or equal to 200 ms as anticipations. Although in the present study using the traditional minimum would have led us to inaccurate conclusions, especially for the older infants, whether the conclusions drawn from other studies would be altered remains an empirical question. Consequently, we recomputed several key analyses from two of our recently published studies for which a higher minimum RT was used (Canfield et al., 1995; Canfield & Smith, 1996).

First, recall that the Canfield et al. (1995) investigation was a longitudinal stability study in which babies viewed an L-R alternating sequence at both 4 and 6 months of age. We reported strong stability of individual differences

123

in median RT ($r = .81$, $p = .001$) but no stability for %ANT ($r = -.12$, N.S.). We also reported that median RT dropped an average of 80 ms during the 4–6-month period but that %ANT did not increase. Instead, %ANT declined from 27% to 18%.

Our reanalysis of these data involved simply recomputing median RT and %ANT using the 133-ms minimum RT and then repeating the original analyses. We found that the effect of reducing the minimum RT was minimal and would in no case have altered the conclusions that we reported. Specifically, the 4–6-month stability increased only slightly for median RT ($r = .87$, $p < .001$), and stability for %ANT increased slightly in the negative direction ($r = -.28$, $p = .18$). The average drop in median RT across the 2-month period remained nearly unchanged (77 ms). When the latencies were classified using the 133-ms minimum RT, the mean level of %ANT was reduced by 4% at both ages (to 23% at 4 months and 14% at 6 months), and the mean of the median RTs was reduced by 15 ms at 4 months and by 11 ms at 6 months.

The Canfield and Smith (1996) study addressed the question of whether 5-month-old infants develop expectations for future stimuli when their occurrence is predicted only by sequential numerical information. The findings supported the interpretation that infants develop number-based expectancies from repeating sets of two items (Study 1) and three items (Study 2). In addition, this report introduced a new dependent variable, chance-corrected anticipation (for a detailed description, see Canfield & Smith, 1996), and average RTs were not reported. Our reanalysis using the 133-ms minimum RT changed neither the overall pattern of findings nor the significance level of any statistical test.

On the basis of these analyses of our previous research findings, we conclude that the practical effect of the minimum RT is insignificant for studies in which babies are 6 months old or younger and also less significant for average RT than for %ANT. It appears that the possibility of a significant discrepancy increases during the 6–12-month period, especially for %ANT. These conclusions are supported by our reanalyses of previously published studies, which revealed that none of the conclusions we reached using the 233-ms minimum need be revised. Using the 133-ms minimum RT had no effect on the stability of individual differences in either average RT or %ANT, and neither did it alter our findings with chance-corrected anticipation. Although we have not analyzed previously published studies involving older infants, concern about the calculation of %ANT in the later months remains. As shown in Figure 26, the difference in %ANT using the two minimums indicates that the magnitude of the difference increases from 5 to 6 months, then again from 8 to 9 months, and yet again from 9 to 12 months. This indicates that a reanalysis of %ANT results from previous studies—especially from the studies of older infants—may alter the original conclusions.

Three studies have reported very noteworthy findings based on VExP data from older infants. Two of these report significant associations between both average RT and %ANT measured at 8 months and either childhood IQ at 3 years or mid-parent IQ (Benson et al., 1993; DiLalla et al., 1990). There is substantial cause for concern that the 8-month-olds in these studies produced large numbers of reactions falling in the 133–233-ms range, leading to an erroneous estimate of %ANT. Moreover, because some of the most important findings of correlations between VExP data and later IQ are based on average *baseline* RTs, which are calculated on five or fewer fixation shifts for each infant, the addition of even one very fast RT could greatly alter the baseline RT value for that child. For example, in our own data, we find that five of 13 8-month-olds (27%) had one or more RTs from among their first five fixation shifts that would have been misclassified as an anticipation had we used the 233-ms minimum.

We also suspect that our minimum RT findings may help clarify the important findings about dimensions of infant information processing reported by Jacobson et al. (1992). Recall that this study included a large number of infant information-processing assessments, including VExP, fixation duration, novelty preference, cross-modal transfer, and other tasks in a study of infants 6.5–12 months old. VExP data were collected only at the 6.5-month assessment. The investigators factor analyzed intercorrelations among the information-processing variables and concluded that there was evidence for two somewhat independent dimensions of information processing, which they labeled *processing speed* and *memory/attention*. An inspection of their factor loadings (see Jacobson et al., 1992, table 5) reveals that, although %ANT has its highest loading (.56) on the memory/attention factor, it also has a substantial loading (−.42) on the processing-speed factor. It is possible that a portion of this loading may arise from the misclassification of very fast RTs as anticipations. Such a measure would include unwanted processing-speed variance and may have been responsible for the curiously high loading of the anticipation variable on the speed factor.

As suggested by our reanalysis of previous studies, we do not expect that conclusions reached by previous investigators using the VExP will be radically altered by using the shorter minimum RT. Instead, we believe that the reduction in error variance would help clarify and strengthen the researchers' original findings. Moreover, an accurate empirical specification of the minimum RT is essential to subsequent explorations of normative age-related change and individual differences in average RT and especially %ANT.

Clarifying what is the relation between %ANT and mean RT is an important issue to be addressed by future research. The major problem seems to be that %ANT is not being measured with acceptable reliability. A possible way to increase the reliability would be to increase the absolute number of anticipations produced. As currently implemented, anticipations in the VExP are rare events. One likely reason is that the stimuli are not highly rewarding and there is little cost associated with not anticipating. The infant's rate of anticipation may reflect a personal cost/benefit analysis, the elements of which vary dramatically from one day to the next. Attempts to increase the reward value of the stimuli and make the reward contingent on anticipatory responses may increase the rate of anticipation and at the same time amplify individual and age differences in the ability to detect the contingencies.

INFANT SACCADE RT: A LINK TO MODELS OF PERFORMANCE AND DEVELOPMENTAL CHANGE

One of the nagging concerns of many infancy researchers is that their research questions, conceptual models, and methods often have closer ties to animal psychology than to theories, models, and methods of human psychology beyond infancy. One need not look far to find contemporary examples of the phenomenon that Haith (1987) labeled the *species gap*. He noted that infancy research can appear as if its object of study is a different species rather than an immature human infant. For example, currently, the most influential model for understanding infant numerical competence is taken, without significant modification, from a model developed to explain counting and timing in the rat (Canfield & Smith, 1996; Meck & Church, 1983). Although we believe that infancy research has benefited from its close association with animal psychology, there remains a concern that the goal of understanding human development from infancy into adulthood can be obscured.

A continuing challenge for researchers is developing conceptual models and methods of study that will lead to an understanding of the nature of cognitive growth from birth to maturity. Thus, infancy researchers are increasingly adapting models of adult attention, learning, and cognition to address questions of infant development. For example, Bhatt and Rovee-Collier (1994) have applied Triesman's (Treisman & Gelade, 1980) model of feature integration in their studies of infant learning and attention. Johnson et al. (1991) have used Posner's (Posner & Peterson, 1990) research and theory on adult visuospatial attention to guide their study of age changes in inhibition of return and other, more general aspects of infant attention. Similarly, Hood and Atkinson (1993) applied Posner's attentional model to study age differences in attentional disengagement during infancy. Finally, when describing their work on infant discrimination learning, Coldren and Colombo (1994) noted the advantages of guiding their research from a model of discrimination learning in children.

We believe the study of development will benefit from this trend. In the remainder of this chapter, we establish links to a developmental model for describing the nature of growth and individual differences in processing speed from childhood to maturity (Hale, 1990; Hale & Jansen, 1994; Kail, 1991b, 1991c, 1993b) and to a general cognitive performance model of processing speed in RT tasks (Grice, 1968; Grice, Nullmeyer, & Spiker, 1977).

Growth in Processing Speed: The Global Trend and Exponential Growth Hypotheses

Models of adult information processing are based largely on measures of the time course of mental processing. RT measures were a key ingredient of the cognitive revolution of the 1950s, and they continue to be the primary measure used to study adult cognition today. Recent research into age changes in the time course of information processing is leading to important generalizations about the nature of cognitive development from childhood into old age. Two related generalizations from these efforts are the global trend hypothesis and the exponential growth hypothesis.

Recall that the global trend hypothesis proposes that, in addition to changes in strategy use and elaboration of the knowledge base, the speed of information processing develops in a uniform manner for all components of the information-processing system (Hale, 1990; Kail, 1993a, 1993b). For example, one prediction from this hypothesis is that the rate of improvement in the speed of mental rotation is the same as the rates of improvement in name retrieval and simple RT, at least from childhood into young adulthood (Hale, 1990; Kail, 1991b, 1991c). The exponential growth hypothesis adds to the global trend hypothesis, stating that the rate at which processing speed develops across all individuals and tasks can be well described by a single parameterization of a simple exponential function that asymptotes in young adulthood (i.e., the strong concept of growth, Kail, 1991a, 1991b; see Chapter VI of this *Monograph*). Thus, processing speed appears to be a dimension of development with the potential to bridge developmental periods from infancy to adulthood, both with respect to normative age trends and as an individual difference variable associated with childhood IQ.

Although researchers have applied adult models to the study of infant information processing, they have not generally investigated individual differences in the shapes of the growth functions or the extent to which this function is similar to or different from the developmental function seen in later phases of development. Because this is a question about the pattern of developmental change, to answer it one's analysis must be focused on understanding the nature of change itself. In this *Monograph,* we described change in terms of individual growth functions for reactive saccade latency, our putative measure of processing speed. Several questions that we explored have important implications for both the global trend and the exponential growth hypotheses.

Because we studied the nature of individual developmental functions, we were able to observe whether the pattern of age change in RT differed for individual infants. Our data show that there is a normative developmental decrease in reaction time during the first year of life. Furthermore, and in general agreement with Kail's (1991b, 1991c, 1993a) research with children and adolescents, we found that this decrease takes the form of an asymptotic exponential function. However, the fact that our individual growth-curve analyses were more consistent with the existence of a local rather than a global asymptote suggests that Kail's parameterization should be bounded regarding the age periods to which it applies. Note that Kail has not claimed that his exponential growth model applies to the description of development outside the age ranges he studied. Nevertheless, our findings suggest that describing growth in processing speed from birth to maturity will require not only different parameterizations of Equation (3) to describe growth for different infants but also at least two parameterizations for different portions of the life span. Additional longitudinal research with toddlers will be needed to test this hypothesis.

Studying individual growth is essential to theory development and testing. Therefore, neither the exponential growth nor the global trend hypothesis can be tested adequately on the basis of cross-sectional data alone. We see no straightforward way to interpret Kail's research unless we make the assumption that group average growth curves, based on cross-sectional data, provide an accurate representation of the nature of growth in each individual. Our data, together with data describing growth in other domains, indicates that this assumption is often unwarranted (Wohlwill, 1973). The problem with cross-sectional data is that one can never determine to what degree averaging data across individuals distorts the shape of the resulting group function. For example, Wohlwill (1973) shows that, when growth takes an exponential form and individuals grow at different rates, simply averaging over individuals can lead to considerable distortion of the true shape of the population function.

We believe that our data are most consistent with some form of the weak concept of development, which allows for substantial individual variation in the pattern of growth. More specifically, they support the view that the form of the developmental function (i.e., exponential) is uniform across individuals and ages but that the parametric values vary across individuals and subgroups. Recall that, on the basis of an inspection of the parameter values of functions estimated for individual infants, we concluded that there were three prototypical functions in our sample (see Figure 16 above). Recall also that the tests of variance estimates obtained from the mixed model suggest significant variation among individuals in the intercept and linear slope. It is important to note, however, that this conclusion clearly depends on whether one considers individual differences in the growth curves to reflect differences in the pattern of growth in various subgroups or only error variance.

Future research should address this important issue. For example, it may be possible to construct a random-regressions model that tests the parameters of the exponential function for significant variation. This type of model would directly address the question of whether individuals are merely randomly distributed about the group function. In addition, greater specificity may be attained by using the clustering techniques described by Burchinal and Appelbaum (1991). The existence of multiple clusters would argue for meaningful individual differences. A sample size larger than 13 will be needed to arrive at firm conclusions about the existence of subgroups. Specifically, a larger sample would have allowed us to judge whether the two infants whose data did not meet the assumptions required by the exponential models were simply prototypical of a common developmental trajectory that is best described by an entirely different family of functions. Finally, the case for noteworthy individual differences could also be explored by determining the behavioral traits or developmental sequelae associated with different patterns of growth in infant RT.

Regardless of whether our data are seen to support the weak concept of development, Kail's research with children and adolescents provided us with a growth model that accurately described age changes in a very different population. It is possible that his specific parameterization of the function represents a second exponential function that applies to growth during childhood and adolescence that asymptotes in early adulthood. If growth during this later developmental period is uniform across individuals, then the strong concept of growth would hold.

However, we caution that additional cross-sectional investigations of group average data will be of no help for deciding on an appropriate growth model for the individual. Similarly, because the global trend hypothesis predicts identical patterns of individual growth on a broad range of speed tasks, it cannot either receive strong support or be falsified on the basis of cross-sectional findings alone. Thus, although Kail's model suggests an appropriate family of functions for individual growth data like ours, his assumptions of a single function with uniform parameters across individuals and tasks have not yet been tested.

Because we concluded that parameters of the growth functions differ for individuals, at least during the first year of life, our data do not support the exponential growth hypothesis, which predicts that a single parameterization should fit all individuals equally well. Furthermore, although we do not believe that the evidence is strong enough to consider it a falsification, recall that the shape of the growth function for ANTL was better described as linear than exponential in our data, suggesting different functional forms for these tasks. Additional research with multiple infant speed tasks will be required to test the hypothesis that, within an individual, the shape of the growth curves for different processing-speed tasks will be identical.

The Variable Criterion Model: What Develops?

Our findings are best interpreted as showing that, in some general sense, speed of information processing develops in a lawful manner during the first year of life. However, just what aspects of processing speed are measured in the VExP is not clear. Greater insight into this issue will depend on formulating and testing models that help specify more clearly the information-processing components that underlie performance. In the context of such a performance model, it then becomes possible to address the question of what develops and how.

As stated earlier, the use of RT measures enables us to apply an adult model of information processing in simple and choice RT tasks to better understand the nature of the performance components. In one approach, the Variable Criterion Model, Grice (1968) begins with the postulate that reactions occur when the strength of excitation for a response exceeds a criterion. The strength of excitation is the result of many independent processes that occur simultaneously and grow monotonically as a function of time since stimulus onset. Therefore, the Variable Criterion Model is one that, besides having other advantages, is more plausible than serial stage models because it is consistent with research showing that motor priming for a response typically begins before the exact target coordinates are encoded (Luce, 1986; Welford, 1980).

There is nothing unique about a performance model that includes an excitation function and a response threshold; what is most distinctive about Grice's model is that the level of the threshold (criterion) is assumed to fluctuate from moment to moment. Grice reasoned that, of the two main components of the model, it was more plausible that the criterion would be time varying and that, because it is a more mechanical process, the rate of growth of sensory excitation is likely to be constant from one moment to the next. Observed trial-to-trial variations in reaction latencies would then reflect variability in the level of the criterion.

It has been demonstrated in studies of adults that the criterion level and variability are influenced by motivation, attention, set, instructions, and sensory adaptation (Grice, Nullmeyer, & Schnizlein, 1979; Grice et al., 1977). Furthermore, it is assumed that the criterion varies from trial to trial according to a normal distribution. Although the normality appears questionable given the shape of RT distributions (see Figure 5 above; see also Luce, 1986), it is not a restrictive assumption and is made entirely for reasons of psychometric convenience (Grice, 1968; Grice et al., 1977, 1982). Finally, the model does not require the assumption of no trial-to-trial variability in the rate of growth of excitatory strength, only that criterion variability is large relative to other sources of variability. Formally, the model treats criterion variability as the only significant source of random variability in RT.

Past research has amassed substantial support for this model. To our knowledge, the relation between criterion variability and age has not been previously discussed, but, as a heuristic model, our data fit it well. When applied to the issue of age-related change in RT, the Variable Criterion Model allows for an intriguing answer to the question of what develops. Under the assumptions of the model, age-related decline in RT results from a reduction in the average criterion level of stimulus information needed to evoke a response. In our data, the model suggests that, as infants get older, their visual system requires a lower level of excitatory strength to initiate a saccade to a peripheral stimulus. One might think of this development as resulting from the combination of experience-dependent and experience-expectant processes (Greenough, Black, & Wallace, 1987). Additionally, the model would suggest that our finding of an age-related decline in SDRT reflects a reduction in the trial-to-trial variability of this criterion level, which leads to increased consistency of responding.

With respect to the question of what develops, the elements of the Variable Criterion Model afford an integration of a number of findings from our own studies with those of other developmental studies. Recall that, on the basis of the pattern of corrective saccades and anticipation error latencies, we concluded that there is little or no age change in the irreducible minimum RT during the 2–12-month period. This intriguing finding is consistent with the notion that the rate of gain of stimulus information changes little from 2 to 12 months. That is, for a given RT trial, when the criterion takes on its lowest value, even the young infant will respond very quickly. This accounts for the occasional 133-ms RT by a 2- or 3-month-old. However, the model posits that the mean RT is determined by the average value of the criterion, implying that our findings reflect the fact that this *average* criterion value, not necessarily the minimum, is much higher in younger babies. When combined with our finding that, independent of the mean RT, trial-to-trial variability (SDRT) declines, our answer to the question of what develops is that the spread and level of the criterion distribution decline during the 2–12-month period. Less stimulus information is needed to evoke a saccade, and the spread of possible criteria is narrowed. What does not appear to undergo much development during this same period is the rate at which sensory-detection information grows following stimulus onset.

Our application of the Variable Criterion Model is depicted graphically in Figure 27. The cumulative normal function represents the growth of excitatory strength for a saccade as a function of time from stimulus onset. On

131

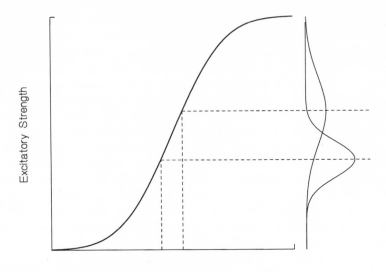

Time from Stimulus Onset

FIGURE 27.—Graphic illustration of the Variable Criterion Model applied to age-related change in infant saccade RT. The function indicates the growth of excitatory strength for responding with a saccade as a function of time since stimulus onset. The normal functions in the right-hand margin represent theoretical distributions of criterion values for two age groups. The older group has less variability and a lower mean level for the criterion, while the criterion distribution for the younger group is more variable and has a higher mean. The figure depicts no age difference in the rate of increase in excitatory strength, but age differences exist for both the means and the variances of the criterion distributions.

the right margin are depicted two normal functions that represent theoretical distributions of criterion values for two age groups. The flatter distribution shows the predicted criterion variability for young infants, and the more peaked distribution represents criterion variability for the older infants. The horizontal dashed lines correspond to the mean levels of excitatory strength for each criterion distribution—emphasizing that the average value of the criterion is lower for the older group. The vertical dashed lines emphasize that this average value is reached sooner after stimulus onset for the older group. Corresponding to our data, this figure uses a single function to represent the rate of increase in excitatory strength, suggesting no age difference in this dimension of performance. In contrast, substantial age differences exist in both the means and the variances of the criterion distributions. Nevertheless, the lower bounds of both criterion distributions are identical, whereas the upper bound is higher for the younger age group.

At first, the conclusions derived from the model appear implausible. Because of the extensive maturation of the fovea, increases in cell size, fiber thickness, and the number of synapses, and the rapid myelination that takes place in the infant visual system during the first 6 months of life (Aslin, 1987; Shea, 1992), one would expect that basic electrochemical properties of the information-processing system would be a major site of development. For example, myelination alone has been estimated to increase nerve conduction velocity by a factor of 30 (Konner, 1991). Furthermore, anatomical constraints have been proposed to have a major effect specifically on the types of processes that influence the rate of accumulation of sensory-detection information in higher centers of the visual system (Aslin, 1987).

However, two other investigations of infant oculomotor control suggest that these sensory factors may play a smaller role in determining optimal performance than they do for average performance. As previously mentioned, Aslin and Salapatek (1975) reported that the saccade latencies of 1- and 2-month-olds were generally quite slow. But Aslin (1987) noted that the mean latency does not tell the whole story of infant competence. Specifically, the shortest latencies of the infants were nearly as short as the shortest latencies of adults who were motivated and trained to respond quickly in a similar paradigm. The minimum RT for infants was 240 ms, compared to 200 ms for adults.

These findings are in agreement with our data regarding the lack of substantial age-related change in the *minimum* time needed for stimulus information to evoke a response, although this minimum would be expected to differ under different stimulus conditions (e.g., stimulus brightness; Grice et al., 1982). Under the assumptions of the Variable Criterion Model, this fact suggests that it is change in the mean level of the criterion that is responsible for the large decline in mean RT from infancy into adulthood, rather than an increase in the rate of accumulation of stimulus information. This is not to say that there is no age-related decline in the minimum, but Aslin and Salapatek's (1975) findings indicate that a decline in the minimum can account for only about 40 ms (less than 10%) of the total age-related change in mean RT from 1 month of age to maturity.

It is also interesting that saccade latencies are not the only realm in which a high level of maturity is found in optimal infant performance. Hainline, Turkel, Abramov, Lemerise, and Harris (1984) report that saccade metrics also show surprising maturity in the very young infant. In a study of saccades in 64 infants, ages 4–151 days, who viewed both simple geometric forms and checkerboards, these researchers found that, when infants scanned the checkerboard, they made a large proportion of saccades that appeared mature. Specifically, the slope of the main sequences (plot of amplitude vs. peak velocity) for even the very young infants did not differ significantly from that of adults. Additionally, no age differences in the slope of the main sequence were observed despite the broad age range. Generally speaking, Hainline et al.'s conclusions would be equally applicable to our findings: "The major conclusion of the present work is that, under some conditions, infants make

133

mature saccades. Using data from infants as young as several weeks who freely scanned visual patterns . . . we found that they could execute saccades whose mean amplitude was similar to that of adults. Their saccades showed regular main sequences. *If we use adult main sequences as a standard, a significant proportion of infant saccades to some stimuli were mature in their execution*" (p. 1777; emphasis added).

The data from Aslin and Salapatek (1975), Hainline et al. (1984), and the present study suggest that development in this domain consists, not of a movement from immature to mature responses, but rather of a shift in the proportion of responses that are mature. There appears to be no stage or developmental period during which adult-like responding is wholly absent. In other words, it may be misleading to view the infant saccadic eye-movement system as uniformly immature. A more accurate view may be that the infant is relatively more *variable* than are adults in executing saccades. The roles of attentional control and motivation would, therefore, appear to loom large in explaining immaturities in infant saccade latencies and metrics (see Hainline et al., 1984).

It is important to note that our application of Grice's model to development does not require the claim that physiological maturation has no influence on age changes in RT. The model requires only that, relative to other sources of variability, the influence of criterion variability is large. Indeed, it is possible to interpret our data as supporting the notion of limited age-related change in the rate of growth of sensory information. Recall that our findings on corrective saccades suggested the possibility that the irreducible minimum RT at 2 months should be placed at 167 ms rather than the 133-ms value that was very clearly indicated for all assessments beyond 2 months (see Figures 7 and 8 above). This may indicate that 2-month-olds do accumulate sensory information at a slightly slower rate than older infants. Pursuing this notion further, when adults produce saccades to partially predictable stimuli, their irreducible minimum RT is generally set at 100 ms (Findlay, 1981; Horrocks & Stark, 1964). These facts indicate that there may be age changes in the rate of growth of stimulus information but that most of this change occurs before an infant's third month. Only a small increase in this type of speed seems to occur between the end of the first year of life and adulthood.

These ideas are not inconsistent with research on the physiological maturation of the oculomotor system. For example, Yakovlev and Lecours (1967; see also Atkinson, 1984; Huttenlocher, 1994) report that myelination of the optic nerve tract, the superior and inferior colliculi, and the optic radiations approaches asymptote when infants are about 3 months old (Konner, 1991). This is consistent with our postulate that most of the age-related change in the rate of growth of sensory-detection information occurs before 3 months.

The first step toward demonstrating the applicability of the Variable Criterion Model is explicitly to model RT density functions, transform them to normal deviates, and obtain their associated theoretical activation functions. Comparisons of parameter estimates from the activation functions among different age groups would address our hypothesis of developmental stability in the rate of growth of sensory-detection information. A second step is to manipulate the variables that have been shown to influence the component processes underlying RT in adults. For example, Jacobs (1993; Nazir & Jacobs, 1992) showed that target discriminability and eccentricity affect processing accuracy, rate, and delay in a visual RT paradigm. This approach can be used with infants if one can succeed in obtaining sufficient numbers of RTs from individual babies.

We mentioned in Chapter I that a possible benefit of developing a paradigm to measure infant RT is that it will allow us to link the facts of infant information processing to the more sophisticated theories and models of child and adult information processing and ultimately help us answer the question of what develops. Our use of Kail's exponential growth model to guide our description of age changes and Grice's Variable Criterion Model to propose an explanation of the age changes (and constancies) in our data represents progress toward this goal. Moreover, our research contributes to the exponential growth hypothesis by analyzing individual growth during infancy. This a priori growth model enabled us to observe differences in the shapes of the individual growth functions and thereby to reject a class of models that assume a strong concept of growth during the first year of life. On the other hand, a great deal of research remains to be done before we have any assurances that the comparisons we have drawn are not altogether too facile. Nevertheless, the point remains that the use of an RT paradigm allows us to pursue a new avenue in the search for similarities and differences between infant information processing and performance and development during later phases of life.

FUTURE DIRECTIONS

There is an unfortunate tradition in both animal and human psychology of conflating the laboratory procedures used to elicit learning with the nature of learning itself. For decades, it was common for psychologists to ask whether a particular instance of learning was truly a case of classical conditioning or operant conditioning. Ethologists studying learning in natural situations have more typically acknowledged the distinction between a paradigm or procedure that is useful for eliciting learning and the nature of the learning process itself (Staddon, 1988). This distinction is important to keep in mind when exploring a new paradigm. In the case of the VExP, we must recognize that its procedures represent only one among many methods that can be used to study infant learning and performance. Nevertheless, the paradigm has notable strengths and the potential to contribute in enduring ways to the science of developmental psychology.

Possibly the most fundamental strength of the paradigm is its use of the oculomotor control system to provide evidence for learning and development. The early maturation of this sensorimotor system enables the researcher to use the same procedure to study infants of widely varying ages. The VExP appears best suited for revealing the development of perception-action linkages and may therefore provide one of the earliest windows to the infant's mind. Although infant reaching begins at around 4 months of age, that is already too late to study many of the important developing brain-behavior relations occurring during the transition out of the newborn period. In addition, because of the tremendous amount of early oculomotor experience, and because moving the eyes involves negligible mass, measures of cognition are less likely to be influenced by purely motoric factors. The habituation paradigm also capitalizes on the infant's visual precocity, but it appears to measure somewhat different features of information processing than the VExP does. The visual habituation paradigm appears to draw more on the processes of object perception and recognition, whereas the VExP likely reflects the perceptual control of actions (Bertenthal, 1996). Thus, the two paradigms complement one another nicely.

By focusing on oculomotor control, VExP researchers can take advantage of the rapidly growing knowledge base in the neural sciences. This represents an important opportunity that is only beginning to be realized. A tremendous amount of knowledge about the neurophysiology of the primate visual system has been generated in the past two decades (e.g., Goldberg & Segraves, 1989; Tusa, 1990). And, although a number of researchers have been exploring brain-behavior linkages in the infant visual system, they have not been much concerned with the neural substrates of anticipatory saccades. Because these saccades are known to be controlled by cells in the frontal eye fields, the VExP may provide a useful tool for studying development in the frontal cortex more generally.

There are already some indications that this may be true. Smith and Canfield (1996) reanalyzed several VExP studies that used 2-month-olds and found that these young babies anticipate appropriately in asymmetrical sequences. We believe the evidence indicates partial functional maturity of the frontal eye fields at a very young age. This conclusion conflicts with Johnson's (1990) model, which postulates that the intracortical connections needed to support anticipatory saccades do not exist until 3–4 months. An important direction for future research is to resolve this apparent discrepancy between VExP research and Johnson's timetable for the functional development of the brain. In addition, future research could use the VExP to address the question of how early visual capacities relate to later reaching indices of frontal cortex development.

Possibly the most important direction for future research with the VExP will be further efforts at construct identification and validation. That is, we must specify more clearly than is currently possible what are the cognitive and developmental processes represented by the various VExP measures. We believe that this will require the use of other paradigms thought to measure the same processes. For example, we need to use multiple measures of information processing to better characterize the processing-speed construct in developmental terms. Additional research should pursue the approach initiated by Jacobson et al. (1992). By including measures from the several major infant paradigms (e.g., various forms of fixation duration and RT, anticipation, novelty preference, operant learning, cross-modal transfer), each administered at several age levels, we may be able to learn how many different dimensions are required to describe infant cognition at particular ages. The use of multiple tasks is also required to address whether age changes in processing speed are specific for individual tasks or whether there is a global exponential trend across development.

Construct validation appears to be most crucial for understanding the anticipation measure. Specifically, it may help us understand why %ANT is a very sensitive index of whether an infant has detected the predictive relations in a sequence of stimuli (e.g., Canfield & Smith, 1996), but it is very insensitive to developmental change over the broad span of infancy. In addition, because the present study found relatively weak age trends and little stability for ANTL, additional research will be needed to show that this variable can be a useful measure of infant information processing.

Again, no single research paradigm can capture all that is interesting or useful to know about infant cognitive development. Progress will depend on increased collaboration among scientists studying similar processes but using different tools. We hope that the research reported in this *Monograph* will encourage such collaborations. And we hope that scientists will be better informed about the implementation and potential value of the VExP as a tool for understanding infant cognitive growth.

REFERENCES

Abrams, R. A., & Jonides, J. (1988). Programming saccadic eye movements. *Journal of Experimental Psychology: Human Perception and Performance, 14,* 428–443.

Appelbaum, M. I., & McCall, R. B. (1983). Design and analysis in developmental psychology. In W. Kessen (Ed.), P. H. Mussen (Series Ed.), *Handbook of child psychology: Vol. 1. History, theory and methods.* New York: Wiley.

Aslin, R. N. (1985). Oculomotor measures of visual development. In G. Gottlieb & N. A. Krasnegor (Eds.), *Measurement of audition and vision in the first year of postnatal life: A methodological overview.* Norwood, NJ: Ablex.

Aslin, R. N. (1987). Anatomical constraints on oculomotor development: Implications for infant perception. In A. Yonas (Ed.), *Perceptual development in infancy: The Minnesota symposia on child psychology* (Vol. 20). Hillsdale, NJ: Erlbaum.

Aslin, R. N., & Salapatek, P. (1975). Saccadic localization of visual targets by the very young human infant. *Perception and Psychophysics, 17,* 293–302.

Aslin, R. N., & Shea, S. L. (1987). The amplitude and angle of saccades to double-step target displacements. *Vision Research, 27,* 1925–1942.

Atkinson, J. (1984). Human visual development over the first 6 months of life: A review and a hypothesis. *Human Neurobiology, 3,* 61–74.

Becker, W. (1972). The control of eye movements in the saccadic system. *Bibliotheca Opthalmoligica, 82,* 223–243.

Benson, J. B., Cherny, S. S., Haith, M. M., & Fulker, D. W. (1993). Rapid assessment of infant predictors of adult IQ: Midtwin-midparent analyses. *Developmental Psychology, 29,* 434–447.

Bertenthal, B. I. (1996). Origins and early development of perception, action, and representation. *Annual Review of Psychology, 47,* 431–459.

Bhatt, R. S., & Rovee-Collier, C. (1994). Perception and 24-hour retention of feature relations in infancy. *Developmental Psychology, 30,* 142–150.

Bogin, B. (1988). *Patterns of human growth.* Cambridge: Cambridge University Press.

Bornstein, M. H. (1985). Habituation of attention as a measure of visual information processing in human infants: Summary, systematization, and synthesis. In G. Gottlieb & N. A. Krasnegor (Eds.), *Measurement of audition and vision in the first year of postnatal life: A methodological overview.* Norwood, NJ: Ablex.

Bornstein, M. H., & Sigman, M. D. (1986). Continuity in mental development from infancy. *Child Development, 57,* 251–274.

Bridges, J. (1990). *GRASP: Graphics Animation Software for Professionals* (Version 3.5) [Computer software]. Ashland, OR: Paul Mace Software.

Bronson, G. W. (1994). Infants' transitions toward adult-like scanning. *Child Development*, **65**, 1243–1261.

Bronstein, A. M., & Kennard, C. (1986). Predictive eye saccades are different from visually triggered saccades. *Vision Research*, **27**, 517–520.

Burchinal, M., & Appelbaum, M. I. (1991). Estimating individual developmental functions: Methods and their assumptions. *Child Development*, **62**, 23–43.

Canfield, R. L. (1988). *Visual anticipation and number perception in early infancy*. Unpublished doctoral dissertation, University of Denver.

Canfield, R. L., & Ceci, S. J. (1992). Integrating learning into a theory of intellectual development. In R. J. Sternberg & C. A. Berg (Eds.), *Intellectual development*. New York: Cambridge University Press.

Canfield, R. L., & Haith, M. M. (1991). Young infants' visual expectations for symmetric and asymmetric stimulus sequences. *Developmental Psychology*, **27**, 198–208.

Canfield, R. L., & Smith, E. G. (1993, March). *Counting in early infancy: Number-based expectations*. Poster presented at the meeting of the Society for Research in Child Development, New Orleans.

Canfield, R. L., & Smith, E. G. (1996). Number-based expectations and sequential enumeration by 5-month-old infants. *Developmental Psychology*, **32**, 269–279.

Canfield, R. L., Wilken, J., & Schmerl, L. (1991, April). *Speed of reaction, expectancies, and mental processing in young children*. Paper presented at the meeting of the Society for Research in Child Development, Seattle.

Canfield, R. L., Wilken, J., Schmerl, L., & Smith, E. G. (1995). Age-related change and stability of individual differences in infant saccade reaction time. *Infant Behavior and Development*, **18**, 351–358.

Carlson, J. S., Jensen, C. M., & Widaman, K. F. (1983). Reaction time, intelligence, and attention. *Intelligence*, **7**, 329–344.

Cattell, R. B. (1957). *Personality and motivation structure and measurement*. New York: World Book.

Cattell, R. B. (1966). The data box: Its ordering of total resources in terms of possible relational systems. In R. B. Cattell (Ed.), *Handbook of multivariate experimental psychology*. Chicago: Rand McNally.

Chambers, J. M., Cleveland, W. S., Kleiner, B., & Tukey, P. A. (1983). *Graphical methods for data analysis*. Pacific Grove, CA: Wadsworth & Brooks/Cole.

Chi, M. T. H. (1977a). Age differences in memory span. *Journal of Experimental Child Psychology*, **23**, 266–281.

Chi, M. T. H. (1977b). Age differences in the speed of processing: A critique. *Developmental Psychology*, **13**, 534–544.

Chi, M. T. H., & Rees, E. (1983). A learning framework for development. *Contributions to Human Development*, **9**, 71–107.

Cohen, L. B., & Gelber, E. R. (1975). Infant visual memory. In L. B. Cohen & P. Salapatek (Eds.), *Infant perception: From sensation to cognition* (Vol. 1). New York: Academic.

Cohen, M. E., & Ross, L. E. (1977). Saccade latency in children and adults: Effects of warning interval and target eccentricity. *Journal of Experimental Child Psychology*, **23**, 539–549.

Coldren, J. T., & Colombo, J. (1994). The nature and processes of preverbal learning: Implications from nine-month-old infants' discrimination problem solving. *Monographs of the Society for Research in Child Development*, **59**(4, Serial No. 241).

Colombo, J. (1993). *Infant cognition: Predicting later intellectual functioning*. London: Sage.

Colombo, J., & Mitchell, D. W. (1990). Individual differences in early visual attention: Fixation time and information processing. In J. Colombo & J. W. Fagan (Eds.), *Individual differences in infancy*. Hillsdale, NJ: Erlbaum.

Colombo, J., Mitchell, D. W., Coldren, J. T., & Freeseman, L. J. (1991). Individual differences in infant visual attention: Are short lookers faster processors or feature processors? *Child Development*, **62**, 1247–1257.

Colombo, J., Mitchell, D. W., O'Brien, M., & Horowitz, F. D. (1987). The stability of visual habituation during the first year of life. *Child Development*, **57**, 474–487.

Cronbach, L. J. (1967). Year-to-year correlations of mental tests: A review of the Hofstaetter analysis. *Child Development*, **38**, 283–289.

Deming, J. (1957). An application of the Gompertz curve to the obscured pattern of growth in length of 48 individual boys and girls during the adolescent cycle of growth. *Human Biology*, **29**, 83–122.

Diggle, P. (1990). *Time series: A biostatistical introduction*. Oxford: Clarendon.

DiLalla, L. F., Thomas, L. A., Plomin, R., Phillips, K., Fagan, J. F. I., Haith, M. M., Cyphers, L. H., & Fulker, D. W. (1990). Infant predictors of preschool and adult IQ: A study of infant twins and their parents. *Developmental Psychology*, **26**, 759–769.

Emmerich, W. (1964). Continuity and stability in early social development. *Child Development*, **35**, 311–332.

Eysenck, H. J. (1986). The theory of intelligence and the psychophysiology of intelligence. In R. J. Sternberg (Ed.), *Advances in the psychology of human intelligence* (Vol. **3**). Hillsdale, NJ: Erlbaum.

Fagan, J. F. (1970). Memory in the infant. *Journal of Experimental Child Psychology*, **9**, 217–226.

Fagan, J. F. (1984). The intelligent infant: Theoretical implications. *Intelligence*, **8**, 1–9.

Fagan, J. F., & Singer, J. (1983). Infant recognition memory as a measure of intelligence. In L. P. Lipsitt (Ed.), *Advances in infancy research* (Vol. **2**). Norwood, NJ: Ablex.

Fagan, J. F. I., Singer, L. T., Montie, J. E., & Shepard, P. A. (1986). Selective screening device for the early detection of normal or delayed cognitive development in infants at risk for later mental retardation. *Pediatrics*, **78**, 1021–1026.

Fantz, R. L. (1961). The origin of form perception. *Scientific American*, **204**, 66–72.

Feingold, A. (1995). The additive effects of differences in central tendency and variability are important in comparisons between groups. *American Psychologist*, **50**, 5–13.

Findlay, J. M. (1981). Spatial and temporal factors in the predictive generation of saccadic eye movements. *Vision Research*, **21**, 347–354.

Findlay, J. M. (1992). Programming of stimulus-elicited saccadic eye movements. In K. Rayner (Ed.), *Eye movements and visual cognition*. New York: Springer.

Fischer, B., & Weber, H. (1993). Express saccades and visual attention. *Behavioral and Brain Sciences*, **16**, 553–610.

Fischer, K. W. (1980). A theory of cognitive development: The control and construction of hierarchies of skills. *Psychological Review*, **87**, 477–531.

Fischer, K. W., & Hogan, A. E. (1989). The big picture for infant development: Levels and variations. In J. L. Lockman & N. L. Hazen (Eds.), *Action in social context: Perspectives on early development*. New York: Plenum.

Fiske, D. W., & Rice, L. (1955). Intra-individual response variability. *Psychological Bulletin*, **52**, 217–250.

Goldberg, M. E., & Segraves, M. A. (1989). The visual and frontal cortices. In R. H. Wurtz & M. E. Goldberg (Eds.), *The neurobiology of saccadic eye movements*. Amsterdam: Elsevier.

Goodkin, F. (1980). The development of mature patterns of head-eye coordination in the human infant. *Early Human Development*, **4**, 373–386.

Greenough, W. T., Black, J. E., & Wallace, C. S. (1987). Experience and brain development. *Child Development*, **58**, 539–559.

Grice, G. R. (1968). Stimulus intensity and response evocation. *Psychological Review*, **75**, 359–373.

Grice, G. R., Nullmeyer, R., & Schnizlein, J. M. (1979). Variable criterion analysis of brightness effects in simple reaction time. *Journal of Experimental Psychology: Human Perception and Performance,* **5,** 303–314.

Grice, G. R., Nullmeyer, R., & Spiker, V. A. (1977). Application of variable criterion theory to choice reaction time. *Perception and Psychophysics,* **22,** 431–449.

Grice, G. R., Nullmeyer, R., & Spiker, V. A. (1982). Human reaction time: Toward a general theory. *Journal of Experimental Psychology: General,* **111,** 135–153.

Groll, S. L., & Ross, L. E. (1982). Saccadic eye movements of children and adults to double-step stimuli. *Developmental Psychology,* **18,** 108–123.

Hainline, L. (1984). Saccades in human infants. In A. G. Gale & F. Johnson (Eds.), *Theoretical and applied aspects of eye movement research.* New York: Elsevier Science.

Hainline, L., Turkel, J., Abramov, I., Lemerise, E., & Harris, C. M. (1984). Characteristics of saccades in human infants. *Vision Research,* **24,** 1771–1780.

Haith, M. M. (1980). *Rules that babies look by.* Hillsdale, NJ: Erlbaum.

Haith, M. M. (1987, June). Expectations and the gratuity of skill acquisition in early infancy. In N. Krasnegor (Chair), *Biobehavioral concepts in development.* Workshop conducted at the National Institutes of Health, Bethesda, MD.

Haith, M. M., Hazan, C., & Goodman, G. S. (1988). Expectation and anticipation of dynamic visual events by 3.5-month-old babies. *Child Development,* **59,** 467–479.

Haith, M. M., & McCarty, M. E. (1990). Stability of visual expectations at 3.0 months of age. *Developmental Psychology,* **26,** 68–74.

Haith, M. M., Wentworth, N., & Canfield, R. L. (1993). The formation of expectations in early infancy. In C. Rovee-Collier & L. P. Lipsitt (Eds.), *Advances in infancy research* (Vol. 8). Norwood, NJ: Ablex.

Hale, S. (1990). A global developmental trend in cognitive processing speed. *Child Development,* **61,** 653–663.

Hale, S., & Jansen, J. (1994). Global processing-time coefficients characterize individual and group differences in cognitive speed. *Psychological Science,* **5,** 384–389.

Hale, S., Myerson, J., Smith, G. A., & Poon, L. W. (1988). Age, variability, and speed: Between-subjects diversity. *Psychology and Aging,* **3,** 401–410.

Hallett, P. E., & Lightstone, A. D. (1976). Saccadic eye movements to flashed targets. *Vision Research,* **16,** 107–114.

Harris, P. L. (1983). Infant cognition. In M. M. Haith & J. J. Campos (Eds.), P. H. Mussen (Series Ed.), *Handbook of child psychology: Vol. 2. Infancy and developmental psychobiology.* New York: Wiley.

Hays, W. L. (1981). *Statistics* (3d ed.). New York: Holt, Rinehart & Winston.

He, P., & Kowler, E. (1989). The role of location probability in the programming of saccades: Implications for "center-of-gravity" tendencies. *Vision Research,* **29,** 1165–1181.

Henderson, C. R., Jr. (1982). Analysis of covariance in the mixed model: Higher-level, nonhomogeneous, and random regressions. *Biometrics,* **38,** 623–640.

Hood, B. M., & Atkinson, J. (1993). Disengaging visual attention in the infant and adult. *Infant Behavior and Development,* **16,** 405–422.

Horowitz, F. D., Paden, L., Bhana, K., & Self, P. (1972). An infant-controlled procedure for studying infant visual fixations. *Developmental Psychology,* **7,** 90.

Horrocks, A., & Stark, L. (1964). *Experiments on error as a function of response time in horizontal eye movements* (Quarterly Progress Rep. No. 72). Cambridge: Massachusetts Institute of Technology.

Howe, M. L., Rabinowitz, M., & Grant, M. J. (1993). On measuring (in)dependence of cognitive processes. *Psychological Bulletin,* **100,** 737–747.

Huttenlocher, P. R. (1994). Synaptogenesis in human cerebral cortex. In G. Dawson & K. W. Fischer (Eds.), *Human behavior and the developing brain.* New York: Guilford.

Jacobs, A. M. (1993). Modeling the effects of visual factors on saccade latency. In G. d'YDewalle & J. Van Rensbergen (Eds.), *Perception and cognition: Advances in eye movement research*. Amsterdam: North-Holland.

Jacobson, S. W., Jacobson, J. L., O'Neill, J. M., Padgett, R. J., Frankowski, J. J., & Bihun, J. T. (1992). Visual expectations and dimensions of infant information processing. *Child Development, 63*, 711–724.

Jensen, A. R. (1982). Reaction time and psychometric "g." In H. J. Eysenck (Ed.), *A model for intelligence*. Berlin: Springer.

Jensen, A. R. (1992a). Commentary: Vehicles of *g*. *Psychological Science, 3*, 275–278.

Jensen, A. R. (1992b). The importance of intraindividual variation in reaction time. *Personality and Individual Differences, 13*, 869–881.

Jensen, A. R. (1993). Why is reaction time correlated with psychometric *g*? *Current Directions in Psychological Science, 2*, 53–56.

Johnson, M. H. (1990). Cortical maturation and the development of visual attention in early infancy. *Journal of Cognitive Neuroscience, 2*, 81–95.

Johnson, M. H., Posner, M. I., & Rothbart, M. K. (1991). Components of visual orienting in early infancy: Contingency learning, anticipatory looking, and disengaging. *Journal of Cognitive Neuroscience, 3*, 335–344.

Johnson, M. H., Posner, M. I., & Rothbart, M. K. (1994). Facilitation of saccades toward a covertly attended location in early infancy. *Psychological Science, 5*, 90–93.

Kail, R. (1988). Developmental functions for speeds of cognitive processes. *Journal of Experimental Child Psychology, 45*, 339–364.

Kail, R. (1991a). Development of processing speed in childhood and adolescence. In H. W. Reese (Ed.), *Advances in child development and behavior* (Vol. 23). New York: Academic.

Kail, R. (1991b). Developmental change in speed of processing during childhood and adolescence. *Psychological Bulletin, 109*, 490–501.

Kail, R. (1991c). Processing time declines exponentially during childhood and adolescence. *Developmental Psychology, 27*, 259–266.

Kail, R. (1993a). Processing time decreases globally at an exponential rate during childhood and adolescence. *Journal of Experimental Child Psychology, 56*, 254–265.

Kail, R. (1993b). The role of a global mechanism in developmental change in speed of processing. In M. L. Howe & R. Pasnak (Eds.), *Emerging themes in cognitive development* (Vol. 1). New York: Springer.

Keppel, G. (1982). *Design and analysis: A researcher's handbook*. Englewood Cliffs, NJ: Prentice-Hall.

Kessen, W. (1960). Research design in the study of developmental problems. In P. H. Mussen (Ed.), *Handbook of research methods in child development*. New York: Wiley.

Kessen, W., Salapatek, P., & Haith, M. M. (1972). The visual response to linear contour. *Journal of Experimental Child Psychology, 13*, 9–20.

Konner, M. (1991). Universals of behavioral development in relation to brain myelination. In K. R. Gibson & A. C. Petersen (Eds.), *Brain maturation and cognitive development: Comparative and cross-cultural perspectives*. New York: DeGruyter.

Larson, G. E., & Alderton, D. L. (1990). Reaction time variability and intelligence: A "worst performance" analysis of individual differences. *Intelligence, 14*, 309–325.

Luce, R. D. (1986). *Response times: Their role in inferring elementary mental organization*. New York: Oxford University Press.

McCall, R. B., Eichorn, D. H., & Hogarty, P. S. (1977). Transitions in early mental development. *Monographs of the Society for Research in Child Development, 42*(3, Serial No. 171).

Meck, W. H., & Church, R. M. (1983). A mode control model of counting and timing processes. *Journal of Experimental Psychology: Animal Behavior Processes, 9*, 320–334.

Merrill, M. (1931). The relationship of individual growth to average growth. *Human Biology, 3*, 37–70.

Michard, A., Tetard, C., & Levy-Schoen, A. (1974). Attente du signal et temps de reaction oculomoteur. *L'Anee Psycholog,* **74,** 387–402.

Miller, J. (1988). A warning about median reaction time. *Journal of Experimental Psychology: Human Perception and Performance,* **14,** 539–543.

Morse, C. K. (1993). Does variability increase with age? An archival study of cognitive measures. *Psychology and Aging,* **8,** 156–164.

Myerson, J., Hale, S., Wagstaff, D., Poon, L. W., & Smith, G. A. (1990). The information-loss model: A mathematical theory of age-related cognitive slowing. *Psychological Review,* **97,** 475–487.

Nazir, T. A., & Jacobs, A. M. (1992). Effects of target discriminability and retinal eccentricity on saccadic latencies: An analysis in terms of variable criterion theory. *Psychological Research/Psychologische Forschung,* **53,** 281.

Nunnally, J. C. (1978). *Psychometric theory.* New York: McGraw-Hill.

Piaget, J. (1952). *Origins of intelligence* (M. Cook, Trans.). New York: Norton.

Poon, L. W., Myerson, J., Hale, S., & Smith, G. A. (1992). *A new look at cognitive speed, variability, and aging.* Unpublished manuscript, University of Georgia.

Posner, M. I., & Peterson, S. E. (1990). The attention system of the human brain. *Annual Review of Neuroscience,* **13,** 25–42.

Ratcliff, R. (1979). Group reaction time distributions and an analysis of distribution statistics. *Psychological Bulletin,* **86,** 446–461.

Ratcliff, R. (1993). Methods for dealing with reaction time outliers. *Psychological Bulletin,* **114,** 510–532.

Regal, D. M., Ashmead, D. H., & Salapatek, P. (1983). The coordination of eye and head movements during early infancy. *Behavior and Brain Research,* **10,** 133–139.

Roberts, R. J., Hayer, C. J., & Heron, C. (1994). Prefrontal cognitive processes: Working memory and inhibition in the antisaccade task. *Journal of Experimental Psychology: General,* **123,** 374–393.

Roberts, R. J., & Ondrejko, M. (1994). Perception, action, and skill: Looking ahead to meet the future. In M. M. Haith, J. B. Benson, R. J. Roberts, & B. F. Pennington (Eds.), *The development of future-oriented processes.* Chicago: University of Chicago Press.

Rose, S. A., & Feldman, J. F. (1987). Infant visual attention: Stability of individual differences from 6 to 8 months. *Developmental Psychology,* **23,** 490–498.

Rose, S. A., Feldman, J. F., & Wallace, I. F. (1988). Individual differences in infants' information processing: Reliability, stability, and prediction. *Child Development,* **59,** 1177–1197.

Ross, R. G., Radant, A. D., Young, D. A., & Hommer, D. W. (1994). Saccadic eye movements in normal children from 8 to 15 years of age: A developmental study of visuospatial attention. *Journal of Autism and Developmental Disorders,* **24,** 413–431.

Roth, C. (1983). Factors affecting developmental changes in the speed of processing. *Journal of Experimental Child Psychology,* **35,** 509–528.

Salapatek, P., Aslin, R. N., Simonson, J., & Pulos, E. (1980). Infant saccadic eye movements to visible and previously visible targets. *Child Development,* **51,** 1090–1094.

Salapatek, P., & Kessen, W. (1966). Visual scanning of triangles by the human newborn. *Journal of Experimental Child Psychology,* **3,** 155–167.

SAS Institute. (1993). SAS for Windows (Version 6.10): SAS PROC MIXED [Computer software]. Cary, NC.

Saslow, M. G. (1967a). Effects of components of displacement-step stimuli upon latency for saccadic eye movement. *Journal of the Optical Society of America,* **57,** 1024–1029.

Saslow, M. G. (1967b). Latency for saccadic eye movement. *Journal of the Optical Society of America,* **57,** 1030–1033.

Shea, S. L. (1992). Eye movements: Developmental aspects. In E. Chelaluk & K. R. Llewellyn (Eds.), *The role of eye movements in perceptual processes.* Amsterdam: Elsevier.

Smith, E. G. (1995). *The presence of predictive saccades in early infancy: A review and revision of existing theory.* Unpublished manuscript, Cornell University.

Smith, E. G., & Canfield, R. L. (1996, April). *Predictive saccades in 2-month-old infants: Implications for current neurophysiological models of visual development.* Poster presented at the meeting of the International Conference on Infant Studies, Providence, RI.

Staddon, J. E. R. (1988). Learning as inference. In R. C. Bolles & M. D. Beecher (Eds.), *Evolution and learning.* Hillsdale, NJ: Erlbaum.

Stark, L. (1971). The control system of versional eye movements. In P. Bach-Y-Rita & C. C. Collins (Eds.), *The control of eye movements.* New York: Academic.

Stark, L., Michael, J. A., & Zuber, B. L. (1969). Saccadic suppression: A product of the saccadic anticipatory signal. In C. R. Evans & T. B. Mulholland (Eds.), *Attention in neurophysiology: An international conference.* London: Appleton-Century-Crofts.

Stigler, J. W., Nusbaum, H. C., & Chalip, L. (1988). Developmental changes in speed of processing: Central limiting mechanism or skill transfer? *Child Development, 59,* 1144–1153.

SYSTAT. (1992). SYSTAT for Windows (Version 5) [Computer software]. Evanston, IL: SYSTAT.

Tanner, J. M. (1962). *Growth and adolescence* (2d ed.). Oxford: Blackwell Scientific.

Tanner, J. M. (1963). The regulation of human growth. *Child Development, 34,* 817–847.

Treisman, A., & Gelade, G. (1980). A feature integration theory of attention. *Cognitive Psychology, 12,* 97–136.

Tusa, R. J. (1990). Saccadic eye movements: Supranuclear control. In R. Daroff & A. Neetens (Eds.), *Neurological organization of ocular movement.* Berkeley, CA: Kugler & Ghedini.

Verhaeghen, P., & Marcoen, A. (1990). Memory aging and the increasing diversity hypothesis: Empirical evidence. *Psychologica Belgica, 30,* 167–176.

Vernon, P. A. (Ed.). (1987). *Speed of information-processing and intelligence.* Norwood: Ablex.

Welford, A. T. (1980). Choice reaction time: Basic concepts. In A. T. Welford (Ed.), *Reaction times.* New York: Academic.

Wentworth, N., & Haith, M. M. (1992). Event-specific expectations of 2- and 3-month-old infants. *Developmental Psychology, 28,* 842–850.

Wheeless, L. L. J., Boynton, R. M., & Cohen, G. H. (1966). Eye-movement responses to step and pulse-step stimuli. *Journal of the Optical Society of America, 56,* 956–960.

Wohlwill, J. F. (1970). Methodology and research strategy in the study of developmental change. In L. R. Goulet & P. B. Baltes (Eds.), *Life-span developmental psychology: Research and theory.* New York: Academic.

Wohlwill, J. F. (1973). *The child psychology series: The study of behavioral development.* New York: Academic.

Yakovlev, P. I., & Lecours, A. R. (1967). The myelinogenetic cycles of regional maturation of the brain. In A. Minkowski (Ed.), *Regional development of the brain in early life.* Oxford: Blackwell.

Young, L. R., & Stark, L. (1962a). *Dependence of accuracy of eye movements on prediction* (Quarterly Progress Rep. No. 67). Cambridge: Massachusetts Institute of Technology.

Young, L. R., & Stark, L. (1962b). *A sampled-data model for eye-tracking movements* (Quarterly Progress Rep. No. 66). Cambridge: Massachusetts Institute of Technology.

Zingale, C. M., & Kowler, E. (1987). Planning sequences of saccades. *Vision Research, 27,* 1327–1341.

ACKNOWLEDGMENTS

We are extremely grateful to the families whose commitment and enthusiasm made this project possible. We recognize Kerri Aaron, Christine Del Favero, Tammy Gotlieb, Linda Kletzkin, Lauren Marder, Lara Pitaro, Nancy Raitano, Natalie Sikka, Emily Vacher, and Becky Wilson for their diligent and expert data collection and processing. We also thank Cindy Hazan, Jeff Haugaard, and three anonymous reviewers for their insightful feedback on a previous version of this *Monograph*. Finally, Ed Frongillo and Charles Henderson provided indispensable statistical consultation. This research was supported by National Institute of Mental Health grant 1-R03-MH45298-01, by Federal Hatch Project 321404, and by the special projects fund of the College of Human Ecology, Cornell University. Portions of this study were presented at the International Conference on Infant Studies, Paris, June 1994, and at the International Conference on Infant Studies, Providence, RI, 1996.

Correspondence concerning this *Monograph* should be directed to Richard L. Canfield at Cornell University, Department of Human Development, Van Rensselaer Hall, Ithaca NY 14850. Electronic mail correspondence may be sent via internet to rlc5@cornell.edu.

COMMENTARY

MODELS OF OCULOMOTOR VARIABILITY IN INFANCY

Richard N. Aslin

Canfield, Smith, Brezsnyak, and Snow have provided a comprehensive exploration of the Visual Expectation Paradigm (VExP) in a small sample of infants between 2 and 12 months of age. The skill and exhaustiveness with which Canfield et al. analyze their data are remarkable. Not only do they focus on a clarification of past studies of the VExP and basic saccadic control in young infants, but they also concentrate on individual growth curves rather than group averages. This is an extremely important aspect of their approach and one that has the potential to reveal details about the intricacies of normative and aberrant development. In general, the Canfield et al. *Monograph* is a tour de force of developmental science, employing longitudinal data, individual growth functions, sophisticated curve fitting, and comprehensive examinations of multiple measures of performance, including often-neglected measures of variability.

Given these positive features of the Canfield et al. *Monograph,* and without trivializing the richness of their data, one can ask what the high points are for the typical reader. The high points are rather simple. First, saccadic reaction time (RT) is the most robust measure of performance, both across age and within infants. The other measures—percentage anticipation (%ANT), standard deviation of RT (SDRT), and anticipation latency (ANTL)—offered little in the way of additional stability or consistency to RT. Second, individual RT growth curves were well fit by polynomial and exponential functions, with exponential functions having more biological plausibility. An evaluation of four models of these RT growth curves suggested that none was clearly superior, although all four models fit the data with correlations greater than 0.93. Third, compelling evidence was provided that the minimum RT (from which %ANT is derived) should be 133 ms (milliseconds) instead of the 200 or 233 ms used in past studies (for the most accessible justification of this new RT minimum, see their Figure 8). However, using the new 133-ms minimum RT made no difference in the conclusions that one would reach in either the present study or previous studies. The sole exception to this conclusion is that the new RT minimum resulted in a slight *decline* in %ANT after 9 months (see their Figure 26). But, because %ANT was inconsistent across age and highly variable within subjects, it is not clear that this decline in %ANT after 9 months has any functional significance.

Perhaps the most intriguing new evidence revealed by the individual growth curves was illustrated in their Figure 17. It appears that some infants show rapid improvements in RT, others show moderate improvements, and still others show little improvement at all between 2 and 12 months of age. These individual differences fuel the speculation that, because it is stable within age (Haith & McCarty, 1990), RT may provide a predictive assay of other cognitive and motor skills during at least the first year of life (Canfield, Wilken, Schmerl, & Smith, 1995) and possibly beyond (Benson, Cherny, Haith, & Fulker, 1993; DiLalla et al., 1990). For RT to become more than a predictor, however, it is necessary to understand what RT means. Does RT reflect a general measure of speed of processing, or is it a more restricted correlate of visuomotor performance, at least partially independent of higher cognitive processing abilities?

To address this question, Canfield et al. correctly note, "Greater insight into this issue will depend on formulating and testing models that help specify more clearly the information-processing components that underlie performance" (p. 130). To this end, Canfield et al. propose that one of two competing models of RT better accounts for the infant VExP data. This model is called the *Variable Criterion Model* because it posits that RT variability results from the growth of neural excitation whose criterion threshold varies from trial to trial. The alternative model is called a *variable rate model* because it posits that RT variability results from the differential rate of growth of neural excitation and a fixed criterion threshold, with that fixed threshold being reached sooner or later on a given trial depending on the different rate of excitatory growth.

Canfield et al. argue that the Variable Criterion Model is consistent with their data. They state that the "age-related decline in RT results from a reduction in the average criterion level of stimulus information needed to evoke a response" and that the "finding of an age-related decline in SDRT reflects a reduction in the trial-to-trial variability of this criterion level" (p. 131). The problem with this interpretation is that these same data are completely consistent with the variable rate model. That is, the age-related decline in RT and SDRT can just as easily be accounted for by an increase in the rate of growth of neural excitation (resulting in faster RT) and a reduction in the variability of neural excitation (resulting in less SDRT).

Canfield et al. also argue that, because it changes little between 2 and 12 months of age, the minimum RT supports the Variable Criterion Model. They state that this "finding is consistent with the notion that the rate of gain of stimulus information changes little from 2 to 12 months" (p. 131). Again, the problem with this interpretation is that the variable rate model is equally consistent with these data. That is, if one simply posits that there is a maximum rate across age by which neural excitation grows after a visual stimulus initiates the programming of a saccade, then the minimum RT will be constant across age. Canfield et al.'s claim that "less stimulus information is needed to evoke a saccade" (p. 131) with increasing age is unsubstantiated. There is no measure of the stimulus information needed to evoke a saccade in any study of infants or of the neural correlates of that saccadic programming across age.

What is needed to definitively choose between the Variable Criterion Model and the variable rate model of RT is a direct assessment of the growth of neural excitation that is triggered by a visual stimulus prior to the initiation of a saccade. A recent study of rhesus monkeys by Hanes and Schall (1996) provides such evidence. Hanes and Schall recorded from single neurons in the frontal eye field (FEF) of the primate cortex. Neurons in the FEF begin to fire approximately 50–150 ms prior to the initiation of a saccade. Of particular interest is that Hanes and Schall found no significant variability in the criterion *threshold* firing of FEF neurons just prior to saccade initiation. They did, however, find a significant and linear relation between saccadic RT and the rate at which FEF neurons increased their firing rate prior to saccade initiation. In other words, saccades always occurred when the FEF neurons reached a particular rate of firing, and the variability in saccadic RT was due to the different slopes by which FEF neurons increased their firing rates from background levels to the fixed threshold level required to initiate a saccade.

Of course, these data from Hanes and Schall are from only one neural region known to be involved in the programming of saccades, and they are taken from adult monkeys. Thus, it is possible that other neural sites (e.g., the superior colliculus) operate according to a different model or that neurons in the FEF may behave differently in early development. Moreover, it is clear that no single neuron is responsible for the programming of saccades; rather, the summed activity of a population of neurons is involved in both the RT and the metrics of saccades. But, at the very least, these data from Hanes and Schall call into question the Variable Criterion Model espoused by Canfield et al., and they are the only definitive neural data available at present to have evaluated directly the Variable Criterion Model and the variable rate model. Finally, it is important to note that the variable rate model is consistent with all the findings reported by Canfield et al.

What then are the implications of the variable rate model supported by Hanes and Schall (1996) for the RT data reported by Canfield et al. and the other studies of the VExP? If the rate of growth of neural excitation, rather than the threshold criterion of neural excitation, accounts for the variability within subjects and across age in saccadic RT, then the average RT in infants is longer than the average RT in adults because of slower rates of growth of neural excitation required to reach a fixed threshold for saccade initiation. The minimum RT, which appears to be developmentally invariant, implies that there is a maximum rate at which neural excitation can grow to threshold. Individual differences in growth functions for saccadic RT imply that some infants have a smaller range of rates of growth of neural excitation, even if minimum RT is constant across infants. Thus, some infants show large decreases with age in average RT, while others show moderate or even minimal decreases with age. The correlation between average RT and later cognitive performance may reflect a more consistent (less variable) growth of neural excitation during the programming of saccades. This consistency may be a correlate of attentional mechanisms that support a wide variety of cognitive tasks, particularly those that require the rapid deployment of concentrated effort.

In conclusion, the Canfield et al. *Monograph* is a superb description of infant saccades and RT performance as well as a prototype for the analysis of individual growth curves during early development. Although some of the underlying models of saccadic programming are subject to alternative interpretations, the Canfield et al. *Monograph* will certainly advance the field of infant development by illustrating the explanatory power of detailed analyses of longitudinal data.

References

Benson, J. B., Cherny, S. S., Haith, M. M., & Fulker, D. W. (1993). Rapid assessment of infant predictors of adult IQ: Midtwin-midparent analyses. *Developmental Psychology, 29,* 434–447.

Canfield, R. L., Wilken, J., Schmerl, L., & Smith, E. G. (1995). Age-related change and stability of individual differences in infant saccade reaction time. *Infant Behavior and Development, 18,* 351–358.

DiLalla, L. F., Thomas, L. A., Plomin, R., Phillips, K., Fagan, J. F. I., Haith, M. M., Cyphers, L. H., & Fulker, D. W. (1990). Infant predictors of preschool and adult IQ: A study of infant twins and their parents. *Developmental Psychology, 26,* 759–769.

Haith, M. M., & McCarty, M. E. (1990). Stability of visual expectations at 3.0 months of age. *Developmental Psychology, 26,* 68–74.

Hanes, D. P., & Schall, J. D. (1996). Neural control of voluntary movement initiation. *Science, 274,* 427–430.

COMMENTARY

INFANT VISUAL EXPECTATIONS: ADVANCES AND ISSUES

Marshall M. Haith, Tara S. Wass, and Scott A. Adler

Despite keen interest over the past three decades in how infants distribute their visual attention, we actually know little about their moment-to-moment processing of the ongoing visual information that typifies an active environment. Since there is reason to believe that the infant's mind manages a flow of information, memory traces, and expectations on a subsecond time scale, our understanding of infant cognition critically depends on procedures and measures that provide a window on this activity. The Canfield, Smith, Brezsnyak, and Snow *Monograph* contributes valuable information about fundamental measures of infant visual processing of an event stream and the underlying processes that they index.

As the authors note, interest in the Visual Expectation Paradigm (VExP) has been growing over the past several years, but the current study is novel and timely in several ways. Much previous work with the VExP has been demonstrational or devoted to issues of reliability or the sensitivity of infants to space, time, and content issues at one or two ages. Developmental issues have not been paramount. Other studies have focused on whether early childhood performance could be predicted from performance in infancy, but sampling during infancy typically consisted of only a single point. This *Monograph* reports the first systematic study of development in this genre of research with repeated sampling at many age points during the first year of life. Simultaneously, it is the first extended longitudinal study of the VExP. The authors have generated a data set that permits examination of a full continuum of reaction time (RT) and anticipatory behavior with exactly the same procedure and materials. Rare, indeed, are such longitudinal data over the first year of life. Happily, the general shape of change in behavior with development fits that of the conglomeration of single-age studies that the authors amassed, many of which used different procedures and materials (see their Figures 1, 2, 10 *a,* and 21 *a*).

Canfield et al. have used this longitudinal set to make original inquiries into individual growth curves and to test various growth models. Longitudinal data permit the assessment of intermonth correlations and, hence, issues of predictability. In addition to extracting full benefit from this approach, they also explored new measures. The study of variance is an important case in point. Researchers who study infants typically focus on mean or median values of performance. Variance is typically conceptualized in terms of differences among individuals that can be attributed to genetic or experiential factors that one cannot control. Usually, intraindividual variance is not an issue because the experimenter has only a single measure, for example, percentage of fixation on a novel stimulus. However, the VExP provides as many as 50–60 measures per condition, so intrasubject variance analyses are possible. Canfield et al. documented for the first time that RT variance declines fairly regularly with age. Anticipation latency is another case in point. Previous studies have reported only percentage of anticipation, providing no information about when the anticipations occurred. Anticipation latency data are helpful in formulating an argument for a criterion break point between anticipations and RTs.

These new findings prompt speculation about underlying process. Consider RT variance. One guess is that intraindividual RT variance reflects fluctuations in infants' attention to the task or, conversely, the ability to sustain attention. This interpretation is consistent with findings and interpretations in the adult literature. The amount of intrasubject RT variability in adults correlates inversely with IQ (Larson & Alderton, 1990). This relation is produced mainly by the larger number of relatively long RTs in individuals who have lower IQs. Our lab has confirmed these findings with infants. At the same time that Canfield et al. were conducting their investigation, Dougherty and Haith (1997) carried out a study that demonstrated that infant RT variability in the VExP at 3.5 months predicted childhood IQ at 4 years of age, with the same inverse relation as reported for adults. There are two points here regarding variance. The general point, suggested by Canfield et al., is that variability in performance can be illuminating and deserves more attention from infancy researchers. The more specific point is that the VExP may be useful in assessing infants' ability to sustain attention in a dynamic task as well as their processing speed and tendency to form expectations.

The basis for age differences in anticipation latency is less clear. The authors suggest that one likely interpretation of an age-related decline in the latency of anticipations is an increase in processing speed, the time it takes for the infant to access an expectancy and initiate an anticipatory eye movement. More discussion of the rationale for this interpretation would have been helpful. In fact, one could as easily predict the opposite because, as infants get older, they might become better at estimating intervals and try to time their anticipation to coincide with picture onset. One study from our lab reported a tendency for even 3.0-month-olds to adjust the timing of their anticipations to the timing of the stimulus, waiting longer to anticipate when the interpicture intervals were longer (Lanthier, Arehart, & Haith, 1993). Latencies did decline up to 9 months of age, a finding consistent with Canfield et al.'s expectations, but they increased for 12-month-olds. While, as the authors suggest, the longer latency in the oldest infants could have reflected lower motivation, it is also possible that the 12-month-olds may have been better than the younger groups at predicting when the next picture would appear.

Our remaining comments concern (1) the dichotomization of reactive and anticipatory behavior; (2) limitations of longitudinal designs; (3) drawbacks in using standard procedures and materials; and (4) inferences that can be made regarding Grice's Variable Criterion Model.

The Dichotomization of Reactive and Anticipatory Behavior

Most of the time it is clear whether an action, in this case an eye movement, is anticipatory or reactive. If an appropriate eye movement occurs prior to picture onset, it could not have been a reaction to that onset. And, if an eye movement occurs as late as 250 milliseconds (ms) after picture onset, there is little argument in the adult or infant literature that the eye movement was a reaction to that stimulus. However, as for all dichotomies, there is room for discussion near the boundaries. The problem arises because, like other actions, eye movements involve a series of processes that precede the observable motion. A decision to act must be made, and that requires time. Then the direction and metrics of the movement must be computed. Finally, the individual uses this computation to execute the movement, a process that involves the time required for neural transmission, neuromuscular excitation, and muscle contraction/relaxation time. For situations similar to those used in the VExP, theorists talk about a sequence involving *disengagement* (of attention from the prior picture location), a *decision* (to move the eye), and *computation* (of direction and distance) for the eye movement (e.g., Becker, 1989; Fischer & Weber, 1993).

The discussion revolves around the interpretation of eye movements that occur within 200 ms after picture onset. Individuals cannot move their eyes instantaneously in response to picture onset, so it is possible that eye movements that occur shortly after picture onset are actually anticipatory. But how do we distinguish between anticipatory and reactive eye movements in the gray zone? An eye movement that occurs, for example, 150 ms after picture onset might represent the result of a computation that was initiated before picture onset, in which case it should be categorized as an anticipation, or it could reflect a reaction to picture onset (i.e., the initiation of the decision stage).

The debate over whether the fastest RT for an infant is 100–133 ms, or 200–233 ms, or somewhere in between, may not seem worth the bother at first glance; the range of discussion is only 100 ms. However, this is a 100% difference between the extreme values, and there are several reasons for getting a fix on this issue: (1) A criterion must be selected for categorizing anticipations and RTs, two types of responding that do not necessarily yield the same results for comparisons between conditions. (2) Stable differences between conditions in the adult RT literature are often of the order of 20–30 ms, and findings in the infancy literature typically involve RT differences of well below 100 ms, so a 100-ms difference in threshold determination is not trivial. (3) Neurophysiological and information-processing models attempt to identify the speed with which independent stages of processing occur (e.g., attentional disengagement, decision making, computation, execution), as assessed by RT; accurate estimates of the fastest RT for infants at different ages are important for modeling how each of these stages develops.

Canfield et al. conducted creative and converging analyses that approached this issue from three directions. The first examined the frequency distribution of eye movements that occurred at varying times around picture onset. There appeared to be a disjunctive rise in frequency toward the RT peak frequency that began at 200 ms following picture onset at 5 months of age but no clear disjunction in younger infants; infants 6 months of age and older showed a frequency rise at +167 ms. These data imply that we should consider any eye movement that occurs as early as 167 ms following picture onset an anticipation for 5-month-olds but adjust the criterion to +133 ms for older infants. The implication for the younger groups is unclear. Researchers should be aware, however, that there are individual differences between infants. Our experience is that some infants in each age group show the group pattern and others do not. Thus, grouped reactions in any time bin in this zone of focus (from +133 to +200 ms) may reflect a mix of anticipatory and reactive saccades. We suspect that, even for an individual infant, some of the eye movements in a time bin may be anticipatory and others reactive.

A second approach to this issue examined the timing of erroneous shifts in eye movement from one side to the other, shifts that could occur only for the irregular (IR) and the left-left-right (L-L-R) sequences, when successive pictures appeared on the same side. The logic here is that error shifts must have been anticipatory (since the next picture actually appeared where the infant was previously fixating), so the latest moment at which they could occur would define the anticipation–reaction time boundary. All erroneous shifts save one occurred prior to the +133-ms bin, which was taken as evidence that the minimal RT was 101–133 ms following stimulus onset and the latest anticipation +100 ms.

Still a third approach examined corrective saccades. Anticipatory eye movements often fall short of the target and are followed by an adjustment eye shift to foveate the target. Such adjustments are less likely if the eye movement is computed from actual stimulus information. Except for the anomalous behavior of 2-month-olds, a break point occurred between +100 ms and +133 ms, with a much higher likelihood of corrections at +100 ms or before than at +133 ms and after. These findings also indicated that eye movements that fell in the +133-ms bin were reactions to the picture, not anticipations.

At first blush, the error and corrective saccade data seem convincing. However, we have our reservations. First, the converging approaches do not lead to a common solution. Whereas the distribution data (approach 1) suggest that the range for the minimal RT is from +167 ms to +200 ms for 5-month-olds, perhaps later for younger infants, and from +133 ms to +167 ms for older infants, the second two approaches suggest that the minimal RT is faster at every age, from +100 ms to +133 ms. This discrepancy is not addressed, but the difference of 25%–50% in criterion is not trivial for models of information processing.

Second, we question the dichotomization of reactions and anticipations in this zone because it is known that preprogrammed eye movements, which would normally play out as saccades, can be affected by stimulus input (Becker, 1989; Becker & Jürgens, 1979). Thus, an eye movement or the *absence* of an eye movement may represent an interplay of anticipatory programming and sensory processing. For example, an eye movement that is triggered to go to a particular location by a flash can be inhibited if a countersignal (e.g., a second visual stimulus that appears at the current fixation location) appears soon enough. The absence of error eye movements (approach 2) occurring 133 ms or more after stimulus onset may represent inhibition by the current picture (at the fixation location) of eye movements that were programmed to move to the wrong side. If so, the data imply that +133 ms may be a threshold for the earliest time that an already programmed eye movement, one in the postdecision, precomputation stage, can be inhibited. A further implication is that correct eye movements that occur at +133 ms, and possibly later, are *real* anticipations (West & Harris, 1993).

A similar criticism holds for approach 3, which examined corrective saccades. Studies have also demonstrated that a saccade that is programmed to move to one location (following a flash) can be recalibrated to move to a farther location if the second flash occurs early enough (Becker & Jürgens, 1979). Thus, an infant who is in the postdecision stage and is computing an anticipatory shift on the basis of his or her memory of the next picture location might be able to adjust the computation on the basis of the stimulus input if the subsequent stimulus occurs early enough before the movement.

In brief, the eye movement may have had its origin in the same memory and expectation processes as "pure" anticipatory eye movements but could also have been influenced by stimulus onset.

The bottom line of this argument is that anticipations and reactions may not be as clearly distinguished as either the Canfield lab or our lab has made them out to be, whether we are referring to the pragmatics of definition for data analysis or the underlying psychological processes. The problem is that, wherever one sets the criterion, there is the possibility of being too liberal on one side and too conservative on the other.

Coincidentally, at the same time that Canfield et al. conducted their longitudinal study, we were carrying out a large-scale cross-sectional developmental study at several of the same ages that they examined (2, 3, 5, and 8 months), with similar picture sequences (i.e., an IR sequence and an L-R sequence, but an L-L-R-R sequence rather than the L-L-R sequence that they used). Although we have not completed our analyses, we can provide some data for purposes of comparison with some of their findings. First, the frequency distributions in latency bins of our data closely matched those of the Canfield et al. data, with a clear rise in frequency at +200 ms, some suggestion of a rise at +167 ms, but no indication of a rise at +133 ms.

Stimulated by the authors' converging analyses, we tried yet another approach. If eye movements that occur at +133, +167, or +200 ms are part of the anticipation distribution, their frequency should correlate more highly with the overall percentage of anticipations (calculated without including eye movements at these values) than with the percentage of fast saccades (those between +233 and +300 ms). On the other hand, if they are part of the RT distribution, the opposite should be true. We found that the percentage of eye shifts in the +133-ms time bin correlated more highly (and significantly) with the percentage of anticipatory saccades than with fast RTs and that those in the +200-ms bin correlated more highly (and significantly) with the percentage of fast RTs than with early shifts. Those at +167 ms correlated more highly with the percentage of fast RTs than with anticipatory saccades, but the correlation was lower than it was at +200 ms.

These findings are for all ages pooled, but there was fair consistency across age. Thus, our inclination is to accept that, indeed, the eye movements in the +200-ms bin are probably reactions (rather than anticipations, as we have held in our prior work), but that those in the +133-ms bin are anticipations, and that those in the +167-ms bin represent a mix. Perhaps eye movements that fall in the +167-ms bin are reactions for some infants and anticipations for others, or perhaps each infant has a mix of reactive and anticipatory saccades that fall within this time bin.

We have two additional reasons for questioning the time values for which Canfield et al. argue. The first is that, with such consistent declines with age in average RT as they obtained, it seems unlikely, although admittedly not impossible, that the fastest RT would remain constant. The second relates to how closely the proposed minimum RT for infants corresponds to that for adults under optimal conditions.

The issue of whether eye movements that occur very soon after stimulus offset represent anticipations or reactions is hotly debated in the literature, not unlike the present discussion (e.g., Fischer & Weber, 1993; Kowler, 1990; West & Harris, 1993). Regardless, no one argues that adults can respond to a stimulus onset in under 80 ms. With current video technology, an eye movement that began even as early as +80 ms would probably end up in the +133-ms bin for reasons that involve peculiarities of VCRs in slow-motion or stop-image mode and video smear as the cue for the detection of an eye movement. Further, adults typically require considerable training and optimal timing parameters to respond within a zone from +100 to +130 ms with what are called "express saccades" (Fischer & Rampsberger, 1984). All these saccades would end up in the +133-ms time bin using current video technology.

Basically, Canfield et al. argue that the fastest RTs of adults under optimal conditions correspond very closely to the fastest RTs of infants as young as 3 months and perhaps as young as 2 months of age under conditions that have not been demonstrated as optimal. Given all the anatomical and neurophysiological changes that occur between infancy and adulthood, we simply have trouble accepting that possibility. Our proposal is that the fastest RT for some subset of the fastest infants probably lies in the region of +150 ms, at least a 50% increment over that of adults, which is usually held to be around +100 ms under optimal conditions.

Limitations of Longitudinal Designs

While longitudinal designs provide unique data and permit unique analyses, the drawbacks should be noted, especially when studying infants. First is the issue of sample size. The Herculean efforts required to obtain many samples from each infant during the first year, combined with finite resources, dictate that only a small number of infants can participate in a given study. Infants' moods are quite variable, the result being that data may be missing for a particular infant at a particular age. The point is that, although the problems involved in data collection are understandable, the resulting sample is still small. In fact, it is amazing that the Canfield et al. data make so much sense given the small sample size, a fact that attests to the care the investigators have taken and the stability of their procedures. Still, the cross-age correlations should be considered cautiously; some involve as few as seven or eight infants in a group. Although there was mixed indication here of month-to-month stability in the percentage of anticipations, two reports from our lab (Arehart, 1995; Haith & McCarty, 1990) indicate stability over at least a 1-week period at 3 months of age. It may be that stability declines beyond detection between an interval of 1 week and 1 month, but it is also possible that there was simply too much variability among too few infants to yield reliable results.

Second, a longitudinal design necessarily confounds task familiarity with age. There is no way to examine, with the present design, the role that multiple exposure to the materials played in the functions that were obtained. However, studies in our lab (Arehart, 1995; Arehart & Haith, 1990) have demonstrated that infants as young as 3 months of age do benefit from prior exposure to the VExP for as long as 1 week. Longer times have not been tested, but older infants are likely to remember aspects of their prior experience better than younger infants, especially with increasing numbers of encounters as they age. Part of the decline in reaction time with age might reflect memory for prior exposure to the VExP, as may the higher cross-month correlations for percentage of anticipation after 7 months of age.

Drawbacks in Using Standard Procedures and Materials

One faces a difficult dilemma in deciding whether to maintain constancy in procedures and materials across age in a developmental study. The advantage of consistency is that one is not vulnerable to the criticism that age differences reflect differences in the experimental protocol. The disadvantages are that memory contamination is more likely as the number of experimental sessions increases and that the same protocol may not be optimal across all ages.

In a task that requires voluntary attention, this issue is particularly acute. We have found that the rate of picture presentation must be increased for older infants in the VExP in comparison to younger infants to maximize their attention and that the pictures must be made more interesting. Canfield et al. acknowledge that infants may not have been as motivated to perform to the same degree at older and younger ages, and we concur that this was a potential problem. RTs might have shown a sharper developmental decline across age if constant motivation could have been maintained, but the issue is especially problematic when evaluating the lack of a developmental increase in anticipatory behavior. We were confronted with this issue in studies with rhesus monkeys using the VExP (Donohue, Braaten, & Haith, 1993). When the VExP was used with monkeys in unaltered form, the monkeys almost never made anticipatory fixations. Several modifications were made to increase their motivation, and the monkeys anticipated over 75% of the time. Thus, the extent to which the procedures motivate the participants can affect the level of anticipatory behavior.

This is an enormous problem. One could spend a lifetime searching for the perfect combination of parameters that equate motivation across such a large fast-growth age range. At times, investigators must simply make a choice and see where it leads. Such is the fate of pioneers. Even in the absence of control for such factors, the developmental data are quite impressive.

What Develops?

An interesting question for this study is whether the ability to form expectations develops. One measure of whether infants form expectations is the occurrence of anticipatory eye movements. No developmental function was obtained for anticipations. Another measure examines the degree to which infant RTs are facilitated; presumably, RTs are faster when the infant forms an expectation for an event than when the event appears unforecasted. Typically, the index of facilitation is a difference score constructed by subtracting a session RT score from a baseline RT score (e.g., RTs to the first five pictures) or by comparing the RTs of a regular with those of an irregular series. There was no discussion of whether the data suggested development in this RT index. A glance at the authors' Table 3, however, suggests that there was no stable change in facilitation across age, at least if one subtracts the RT for the L-R or L-L-R series from that for the IR series. It would have been helpful had the authors explicitly discussed the issue of whether the ability to form expectations changes with age. Of course, it does, but there appear to be no confirming data here. Again, the difficulty in maintaining constant motivation across age may be a key factor.

Canfield et al. focus instead on the factors that may control RT change with age, their most stable finding. One issue that they address is Kail's strong interpretation of development as opposed to a weaker version. However, the present study was not really designed to address this issue because information about performance on a variety of tasks is needed. Although, as Kail predicts, a single coefficient did not seem to describe the current infant data, because of limitations imposed by sample size, missing data, and outriders, the data do not provide an adequate test of Kail's proposal. Still, the modeling effort that the authors undertook serves as a nice example of questions and approaches that one can address with their techniques.

Finally, the authors suggest that their data fit Grice's notion that variation in RT reflects a change in the criterion that subjects use to trigger an action, with less sensory information required at older than at younger ages. However, we are not convinced that the alternative interpretation should be shelved; that is, the criterion could remain relatively constant across age, with variation reflecting differences in the rate at which activation (from sensory information) accumulates. Canfield et al.'s argument is based on the lack of change of the "irreducible minimum" RT across age, an assumption that we questioned above.

A recent study prompts further skepticism about Grice's Variable Criterion Model. Hanes and Schall (1996) addressed this issue in a study of rhesus monkeys in which they simultaneously recorded eye movements in an RT task and single neurons in the frontal eye fields. Neurons in the frontal eye fields are involved in voluntary eye movements and increase their activity about 100 ms before an eye movement to a peripheral target in their movement field. Hanes and Schall tested Grice's Variable Criterion Model against a variable rate (of accumulation of activation) model to account for RT variance by measuring the level of neural activity 15 ms before each eye movement. The Variable Criterion Model predicts that the level of activity just before an eye movement will vary as the RT varies, whereas a variable rate model predicts that neural activity will be approximately constant before all RTs; what should vary with RT is the time it takes for activity to reach the premovement level.

The data impressively supported a variable rate model. Neuronal activity levels remained virtually constant before eye-movement RTs that varied from around 215 ms to 310 ms. There was also a close linear relation between variation in RT and variation in the rate of growth of activation to the same common level just before eye movement. Of course, we do not have comparable data for infants to test whether intraindividual variation in RT or developmental changes in RT also reflect variation in the rate of accumulation of sensory activity. It is possible that the Variable Criterion Model could account for this variation early in life but not later. In either case, an important question is what psychological factors make the critical factor vary, whether it be the criterion or the rate of accumulation of sensory activity. A likely hypothesis is that variations in the intensity and/or maintenance of attention are involved, but whether this is so and the specific routes by which these influences work await further resolution.

In sum, the Canfield et al. *Monograph* adds valuable information to the literature about RT and anticipation performance and variability of performance in the VExP. There are lessons to be learned from reading this *Monograph* about the value of longitudinal data and how longitudinal data can be uniquely exploited to address interesting developmental issues. Novel questions were raised, and models were proposed regarding infant performance in dynamic visual tasks that provide stimulating bases for contending views. This work also serves as a prototype of thoughtfulness and justification of decisions for design and data analysis; rarely are the thought processes behind such decisions provided for the reader. We noted some concerns, which, of course, one has with any study. However, this *Monograph* fills a large gap in the literature and raises many questions that are new to the field. We look forward to further exploration by infant researchers of the trails that this *Monograph* has broken.

159

References

Arehart, D. M. (1995). *Young infants' formation of visual expectations: The role of memory, rule complexity, and rule change.* Unpublished doctoral dissertation, University of Denver.

Arehart, D. M., & Haith, M. M. (1990, April). *Memory for space-time rules in the infant Visual Expectation Paradigm.* Poster presented at the International Conference on Infant Studies, Montreal.

Becker, W. (1989). Metrics. In R. H. Wurtz & M. E. Goldberg (Eds.), *The neurobiology of saccadic eye movements.* New York: Elsevier.

Becker, W., & Jürgens, R. (1979). An analysis of the saccadic system by means of double step stimuli. *Vision Research,* **19,** 967–983.

Donohue, R. L., Braaten, R. F., & Haith, M. M. (1993, March). *An animal model for the formation of visual expectations in human infants.* Poster presented at the meeting of the Society for Research in Child Development, New Orleans.

Dougherty, T. M., & Haith, M. M. (1997). Infant expectations and reaction time as predictors of childhood speed of processing and IQ. *Developmental Psychology,* **33,** 146–155.

Fischer, B., & Rampsberger, E. (1984). Human express saccades: Extremely short reaction times of goal directed eye movements. *Experimental Brain Research,* **57,** 191–195.

Fischer, B., & Weber, H. (1993). Express saccades and visual attention. *Behavioral and Brain Sciences,* **16,** 553–610.

Haith, M. M., & McCarty, M. (1990). Stability of visual expectations at 3.0 months of age. *Developmental Psychology,* **26,** 68–74.

Hanes, D. P., & Schall, J. D. (1996). Neural control of voluntary movement initiation. *Science,* **274,** 427–430.

Kowler, E. (1990). The role of visual and cognitive processes in the control of eye movement. In E. Kowler (Ed.), *Reviews of oculomotor research: 4. Eye movements and their role in the cognitive processes.* London: Elsevier.

Lanthier, E. C., Arehart, D., & Haith, M. M. (1993, March). *Infants' performance with a nonsymmetric timing sequence in the Visual Expectation Paradigm.* Poster presented at the meeting of the Society for Research in Child Development, New Orleans.

Larson, G. E., & Alderton, D. L. (1990). Reaction time variability and intelligence: A "worst performance" analysis of individual differences. *Intelligence,* **14,** 309–325.

West, P., & Harris, C. M. (1993). Are express saccades anticipatory? *Behavioral and Brain Sciences,* **16,** 593–594.

CONTRIBUTORS

Richard L. Canfield (Ph.D. 1988, University of Denver) is an associate professor in the Department of Human Development at Cornell University. His current research focuses on information-processing models of cognitive development, infant predictors of childhood cognitive competence, the development of number concepts, and the relation between low-level lead exposure and early cognitive development.

Elliott G. Smith (M.S. 1991, Villanova University) is a doctoral candidate in the Department of Human Development and Family Studies at Cornell University. His current research interests include neuroanatomical development and its relation to early anticipatory abilities, effects of early lead exposure on later cognitive abilities, and numerical competence of young infants.

Michael P. Brezsnyak (B.A. 1994, Cornell University) is a graduate student in clinical psychology at the University of Colorado at Boulder. His research interests include bipolar treatment outcome, social relationships and their effects on the course of psychopathology, infant cognition, and the use of computers in psychological research.

Kyle L. Snow (M.S. 1994, Cornell University) is a visiting assistant professor of psychology at Wilkes University. His research has focused on transactional models of mother-infant interaction and relations between social interaction and cognitive development during infancy.

Richard N. Aslin (Ph.D. 1975, University of Minnesota) is professor of brain and cognitive sciences at the University of Rochester. His research spans a variety of domains within the study of human infants, including basic sensory abilities, oculomotor control, object perception, and speech and language processing. His most recent work has been directed toward the role of attention in eye movement control and to studies of the statistical learning of sound patterns in artificial languages.

161

Marshall M. Haith (Ph.D. 1964, University of California, Los Angeles) is a professor of psychology at the University of Denver, where he holds the John Evans Chair. He is also a clinical professor of psychiatry at the University of Colorado Health Sciences Center and a National Institute of Mental Health research scientist. His research interests have been in the areas of perceptual and cognitive development in infants and children and have focused most recently on the development of visual expectations in infants.

Tara S. Wass (M.A. 1997, University of Denver) is a National Institute of Alcohol Abuse and Alcoholism predoctoral fellow in developmental psychology at the University of Denver. She is interested in cognitive and neural development during early infancy. Her research program focuses on the processes underlying anticipatory and reactive eye movements in normal and fetal-alcohol-exposed infants.

Scott A. Adler (Ph.D. 1995, Rutgers University) is a National Institute of Mental Health postdoctoral fellow in developmental psychology at the University of Denver. His graduate research program focused on perceptual processes, learning, and memory in infants. He is currently examining the roles that content and timing play in infant visual expectations. He received the 1996 Outstanding Dissertation Award from the International Society on Infant Studies.

STATEMENT OF EDITORIAL POLICY

The *Monographs* series is intended as an outlet for major reports of developmental research that generate authoritative new findings and use these to foster a fresh and/or better-integrated perspective on some conceptually significant issue or controversy. Submissions from programmatic research projects are particularly welcome; these may consist of individually or group-authored reports of findings from some single large-scale investigation or of a sequence of experiments centering on some particular question. Multiauthored sets of independent studies that center on the same underlying question can also be appropriate; a critical requirement in such instances is that the various authors address common issues and that the contribution arising from the set as a whole be both unique and substantial. In essence, irrespective of how it may be framed, any work that contributes significant data and/or extends developmental thinking will be taken under editorial consideration.

Submissions should contain a minimum of 80 manuscript pages (including tables and references); the upper limit of 150–175 pages is much more flexible (please submit four copies; a copy of every submission and associated correspondence is deposited eventually in the archives of the SRCD). Neither membership in the Society for Research in Child Development nor affiliation with the academic discipline of psychology are relevant; the significance of the work in extending developmental theory and in contributing new empirical information is by far the most crucial consideration. Because the aim of the series is not only to advance knowledge on specialized topics but also to enhance cross-fertilization among disciplines or subfields, it is important that the links between the specific issues under study and larger questions relating to developmental processes emerge as clearly to the general reader as to specialists on the given topic.

Potential authors who may be unsure whether the manuscript they are planning would make an appropriate submission are invited to draft an outline of what they propose and send it to the Editor for assessment. This mechanism, as well as a more detailed description of all editorial policies, evaluation processes, and format requirements, is given in the "Guidelines for the Preparation of *Monographs* Submissions," which can be obtained by writing to the Editor, Rachel K. Clifton, Department of Psychology, University of Massachusetts, Amherst MA 01003.